Analytical and Computer Cartography

**Prentice Hall Series in
Geographic Information Science**

Analytical and Computer Cartography

2nd Edition

Keith C. Clarke

Hunter College—City University of New York

Prentice Hall, Englewood Cliffs, New Jersey 07632

Library of Congress Cataloging-in-Publication Data

Clarke, Keith C., 1955-
    Analytical and computer cartography / Keith C. Clarke–2nd ed.
        p.   cm.
    Includes bibliographical references and index.
    ISBN 0-13-033481-2
    1. Cartography–Data Processing.   I. Title
GA102.4.E4C377 1995
526'.0285--dc20   94-33245
                              CIP

Acquisitions editor: Ray Henderson
Editorial/production supervision: Kathleen M. Lafferty/Roaring Mountain Editorial Services
Copy editor: Stephen C. Hopkins
Cover design: Pat Woczyk
Manufacturing buyer: Trudy Pisciotti

 © 1995, 1990 by Prentice-Hall, Inc.
A Simon & Schuster Company
Englewood Cliffs, New Jersey 07632

Printed in the United States of America

10 9 8 7 6 5 4 3 2 1

ISBN 0-13-033481-2

Prentice-Hall International (UK) Limited, *London*
Prentice-Hall of Australia Pty. Limited, *Sydney*
Prentice-Hall Canada Inc., *Toronto*
Prentice-Hall Hispanoamericana, S.A., *Mexico*
Prentice-Hall of India Private Limited, *New Delhi*
Prentice-Hall of Japan, Inc., *Tokyo*
Simon & Schuster Asia Pte. Ltd., *Singapore*
Editora Prentice-Hall do Brasil, Ltda., *Rio de Janeiro*

For those I have taught,
for those I will teach,
and to those who taught me.

Contents

10   A Transformational View of Cartography   171

11 Map Transformations   183

12 Data Structure Transformations   226

13 Terrain Analysis   249

Foreword

In the twentieth century, cartographers have seen the adoption of the metric system, designation of Greenwich as the prime meridian, specification of an international sequence of colors for terrain maps, the beginning of an international millionth map of the world, the almost universal adoption of plane coordinate systems for local surveys, and the list goes on. But one is not allowed to write such long sentences. Still it is necessary to mention automobiles and road maps, airplanes and aeronautical charts, and aerial photography and photogrammetry. These latter items have been extended to satellite altitudes and to wavelengths—radar and infrared—beyond human sensing capabilities. The electronic revolution has led to the replacement of triangulation by trilateration, and of course, to computers.

The statistical information collection agencies have kept up, providing more data for thematic maps and atlases, until recently scribed on modern stable plastics, but increasingly generated by computer graphics software. Getting the world into the computer is still a problem, but more and larger databases are fast becoming available. For those who know how to use it, there has never been available nearly as much knowledge about geography.

This is true at the local, regional, and international level. The rapidly growing geographic information systems contain formalizations of cartographic concepts that have evolved over thousands of years. The topological structuring of geographic data is so basic that it almost had to be rediscovered. With more and more of the solutions to everyday geographical problems solvable by algorithmic procedures—minimal path routines combined with a digital street network stored in an automobile provide one currently implementable example—those tasks that humans perform more quickly still rely on the impressive processing capabilities of the visual cortex. Thus modern cartography requires both the artistic for geographic illustration and the analytic for solving geographic problems. The introductory period of computer-assisted cartography dealt with mathematically simple, but tedious, operations: survey adjustments, ellipsoidal calculations, map projections, replacement of manual drafting by plotters. More difficult tasks are now being approached—for example, map name placement or map generalization—using systems of expert rules which attempt to mimic the intuitive heuristics used by well-trained cartographers.

All this, and more, is the domain of analytical and computer cartography.

Waldo R. Tobler

Preface

Since I wrote *Analytical and Computer Cartography,* nothing has become more clear than the important role that analytical cartography has to play in the years to come. Analytical cartography has become the solid bedrock of theory upon which the field of geographic information systems is rapidly building. Technical language and concepts from analytical cartography are already entering basic texts and even the broad public consciousness. Every day, exciting new applications of both the science and technology of cartography are making tangible improvements to society at large. At the national level, cartographic and geographic data now seem to be considered as the "trucks" that will pass along the national information superhighway during the new information age. As yet, both the academic and the applied fields show no end to growth. It is the task of cartographers to keep the principles sufficiently in touch with the applications so that serious errors and misapplications are avoided during this period and so that the science works for the public good.

This edition is far more than a minor revision. In switching word processors, I have redrafted every single graphic in the book, in many cases by capturing output from computer programs directly into the text. This capability barely existed when I wrote the first edition in 1989. I have added two entirely new chapters, one on using the new information resources available on the Internet to find cartographic data and software, and one on map design. The latter has come about from the dozens of times I have been asked to cover this material as part of this book. I have included the programs from the text, in most cases rewritten as functions, on a disk with the book, because I was asked many times for this by readers of the first edition. I have made many other improvements, including dividing up the longer chapters and by making a few corrections, for which I thank the wonderful people who have taken the time to write and tell me where they had found errors in the first edition.

Michael Goodchild has labeled much of the content of this book *Geographic Information Science,* a thoughtful choice of names which has much to offer academic consideration. This revision launches a Prentice Hall series with this new field as a focus, one which I hope will continue to serve the discipline for years to come. I urge readers to seek out the books in the series, which will add depth and breadth around a series of core or basic books.

Once again, thanks are due to many. Kathleen Lafferty proved to be an exacting and good-humored production editor, and the final form of this edition owes much to her. In addition to the reviewers named in the first edition, reviews were conducted by Douglas Banting, Nick Chrisman, Jeremy Crampton, David Lanter, and Mike Peterson. Thanks to Verislav Tesovic for figures in Chapter 13; to Cheryl Weisner for assistance in research; to Diming Chen and Jason Gresh who contributed computer code; and to the students in GTECH 380, 722, and 731 at Hunter College who continue to find mistakes in the text. Special thanks go to Ray Henderson at Prentice Hall, who has served as instructor, climbing companion, and friend during my writing endeavors and whose support and encouragement have brought into being the *Prentice Hall Series in Geographic Information Science.*

Last, but most importantly of all, I thank my wife, Margot, and my daughters, Chantal, Elizabeth, Anne, and Caroline. Their love, laughter, tolerance, and support make all this possible and give it purpose.

Keith C. Clarke

Analytical and Computer Cartography

PART I
Computer Cartography

This section of three chapters introduces the fields of *analytical and computer cartography* and points out that computer cartography deals primarily with the tools of the current technology with which maps are produced. Chapter 1 gives a brief history of computer cartography and examines some of the impacts that the computer has had upon the discipline. General trends in cartography as a result of the computer are considered, including what advantages and disadvantages the the use of the computer has introduced. In addition, the link between computer cartography and geographic information systems is placed in context.

Chapters 2 and 3 introduce the concepts behind, and the variety of, software and hardware environments for computer mapping. The concept of a hardware device, as a specialist item for graphic input and output, is central to the discussion. The cartographic workstation is developed as a generic workplace for the cartographer. Software is placed in historical context, and the reader is pointed to other information resources from which to gain useful information on software reviews and acquisition.

1

The Computer and Cartography

1.1 THE COMPUTER AT THE CENTER OF CARTOGRAPHY

Cartography is a discipline as old as humankind and as young as today's newspaper. Why is this? Cartography is old because as a means of expression the map probably predates many other forms of human communication, and maps survive that are several thousand years old. Cartography is young because it is a discipline that has been subjected to a series of revolutions in innovative technology. At first, these revolutions came separated by centuries, and in the case of the era following the decline of Greek and Roman cartography most of the cumulative knowledge of mapmaking was forgotten. Since the advent of the *digital computer,* however, we have become used to cartography as being in a state of almost constant technological revolution.

Why has this technological tool surpassed all others in the history of the discipline? How have cartographers adapted their discipline to this constant state of flux? And what will cartography look like in the years of innovation to come? In this book, we examine the computer revolution in cartography from the standpoint of how cartography has changed, and more important from the standpoint of how it has remained the same. As we shall see, the revolution has shaped a new cartography, in which the specifics of technology appear to be overwhelming.

Fortunately, the revolution has also sent cartography back to its historical and mathematical roots and therefore has made the basic principles of cartography as much in demand, if not more so, as at any previous time. When we think of the technology of mapmaking, the center of activity now lies within computer cartography. In the past, we thought of cartographers as scribes with quill pens scratching out maps of the world. This is simply no longer the way it is done. Mapmaking technologies have come and gone, and the simple origins are now very distant. Cartographers have changed the way they see the creation of maps and the role of mapmaking itself within cartography. This has been the case in both manual and computer cartography, for cartography is a set of skills and a body of theory, and the theory remains the same independent of what particular technology one happens to use to make any particular map.

This is why this book has the title *Analytical and Computer Cartography,* for there are two interlinked themes. The first theme, *analytical cartography* (Tobler, 1976), deals with the theoretical and mathematical background behind cartography and the rules cartographers employ in the mapping process. The other theme is hands-on *computer cartography,* the particular set of methods and techniques that the current technology uses to produce maps. If we use the term *computer cartography,* then in the past we have had quill-pen cartography, mapping-pen cartography, scribing cartography, and photogrammetric cartography, for these are some of the previous technologies we have used in the science of mapmaking. This book instructs the reader in mapmaking using the most current tool, the digital computer.

This does not necessarily affect the body of theory behind cartography, but rather emphasizes and reemphasizes the cartographic lessons of the past. For example, for most map projections, the underlying equations and transformational geometry were worked out, sometimes to perfection, in previous cartographic eras. Today the equations still work whether we produce the map by hand construction or by computer graphics. A typical laboratory for computer cartography may contain microcomputers with graphics cards, high-resolution color monitors, digitizers, color plotters, and printers rather than drafting tables and sinks. Although a student in a class on computer cartography may use such a laboratory now, we probably (in fact definitely) will have something different just a few years from now. This is the nature of scientific and technological revolutions. Most of the theory and principles presented and reviewed in this book will work just as well using the technology of next year or the next decade, however, just as most of the principles worked when they were derived centuries ago.

1.2 STAGES OF ADAPTATION OF THE COMPUTER

Morrison (1980) stated that there are three stages in the adaptation of a new technology. First of all, we have a reluctance to use the new technology; we close our eyes and pretend it isn't there. For example, in word processing technology, we may characterize this stage with statements such as, "I've been using an XYZ brand typewriter for 20 years, and it always works just fine. What are these word processors? I really don't need one, they are just too expensive." This is the *reluctance to use* stage. In the second stage, the *replication* stage, technology attempts to replicate the previous technology. If we return to the typewriter example, we might find IBM replacing its Selectric with a "memory" typewriter, with a single line display and some limited editing capabilities. This uses just a little of the new technology but does not "embrace" it, that is, fully take advantage of all of its features to do new things. We are still using the old technology, but are simply making it slightly better. We are copying the way typing was always done, that is, line by line.

In cartography, computer cartography was ignored for some time, as in the first stage, and then was faced by replication, producing pen plotters and table digitizers, which were simply updated electronic versions of the pens and drafting tables we had always used. The plotters were simply mechanical arms with pens fixed to them, and we still fed them pieces of paper and changed their ink. We took a digitizing tablet, much like a drafting table, and instead of using a mapping pen we used a cursor to draw lines. We replicated the previous technology, making maps exactly the way we used to, using only some aspects of the new technology.

The third stage is the full implementation of the new technology, in which we forget the previous technology, and the new technology becomes the current technology. Cartographers first had a reluctance to use computers altogether ("No computer can draw a map the way I can"), then they replicated the previous technology ("Well, maybe those computers can draw maps after all, but let me see them draw on Mylar and Leroy stencil the way I can"). Finally, in the full implementation of the technology we have to ask new questions altogether. This is a sign that a revolution has taken place, because new ideas are necessary to organize the new approach. We may ask, for example, if maps actually have to be on paper. New media are now available, such as microfilm, video, broadcast images, and holograms. Perhaps, more simply, we now have to ask ourselves just what a map actually is.

A characteristic of a developing technology is that we move through these stages. Also, however, we find that all stages usually exist at the same time. We still have both an ignorance of and a reluctance to use this technology. We have replication of the previous technology, but fortunately we also have some full implementations, and there are some good examples of people who have completely adopted computer technology and made a success of it.

More recently, Morrison (1993) has argued that the impact of the computer has gone even further and is eroding the boundaries of the organizations that conduct and control cartography. The flexibility of computer networks, the increasing collaboration between universities, government and industry, and the new highly portable technologies have made the formal cartographic organizations at least partially obsolete. Such institutional changes are likely to significantly impact the way cartographic data, expertise, and mapping problems are assembled to make maps with future mapping systems. The inevitable result will be a more spatially literate population making use of the tools and techniques of cartography in new and exciting ways. Chapter 6 provides an introduction to the critical issues of the impact of networks on cartography.

1.3 THE HISTORY OF COMPUTER CARTOGRAPHY

Computer cartography in the United States really dates back to a single article written by a graduate student at the University of Washington, Waldo Tobler. The paper "Automation and Cartography," was published in the *Geographical Review* in 1959. At that time, plotting devices were simple cathode-ray tubes, and input and output were normally by punched cards. During the early years, the 1960s, the accent was on the creation of algorithms, that is, the creation of expressions of ways of doing things mechanically that had previously been done by hand. Because programming these algorithms was difficult, many remained unimplemented. So in the past, contour lines had been drawn by a cartographer, using knowledge about the lay of the land and perceptions about how the map should look. Later, the computer gained this ability, giving the cartographer the role of deciding how best to represent the lay of the land rather than how to draw contour lines.

The first problem that had to be solved was how to make a computer draw cartographic lines. The way to do that was to examine how the cartographer had made the decision, what methods were in use, and how they could be automated. In fact, many of those decisions are made on a very simple mathematical basis, and there is often a simple algorithm that will replace the method the cartographer has been using.

Cartographers devised a plethora of different algorithms, at first implemented as stand-alone computer programs and later consisting of program packages, peaking with the production in 1968 of the SYMAP package at Harvard University. The 1970s saw two major changes. First of all, the implementation of new algorithms brought innovations in producing new types of maps that had previously been impossible, certainly computationally, to produce. Within five years, most cartographic techniques were automated, even some complicated methods such as hill shading and cartograms. Computer programs were created that could produce almost all the various types of cartographic representation.

The second change, during the 1970s, was that people began to realize that applications of computer cartography had potential commercial value, just as a whole first generation of graduates from the universities that were specializing in computer cartography provided a small group of trained students for the job market. Many small companies started during this period, employing the new cartographers and using software engineering practices to produce some really effective and cost-competitive cartographic software. Most of this software was made available for those who wished to make new maps or those who wished to replace manual mapping systems. Federal, state, and local government frequently took the initiative in producing mapping software. This phase continues today and in fact is gaining momentum as the industry matures and as the microcomputer and workstation penetrate smaller and smaller drafting-office environments.

Some of the first programs were for the types of computers that were around in the 1970s: large mainframe computers, mostly IBM products that used existing languages such as FORTRAN and simple proprietary graphics plotting systems such as CALCOMP and Tektronics PLOT-10. In keeping with the times, most applications programs ran in batch mode and frequently used the line printer as the primary display device. A more detailed discussion of specific software packages and computer programs follows in Chapter 3.

1.4 NEW DISPLAY MEDIA

Computer cartography has dictated the development of new types of hardware. At the same time, computers have become much smaller and cheaper; in fact, we now use computers on our laps. Storage ability has improved greatly, bringing down the need to force things to fit into the smallest possible space. We have whole new storage technologies that did not exist in the early 1980s: optical discs, video, bubble, 8-mm tape and so forth. The CD-ROM in particular has become a major medium for the storage and distribution of map data. The four-disc set produced by the Defense Mapping Agency called the Digital Chart of the World, for example, contains 14 data layers including vegetation, coastlines, roads, hydrology, and place names for the entire world at a scale of 1 to 1 million.

We have new methods and new display media, which have broadened our minds about what a map is. We used to think of a map as something we could roll up in a tube. Now a map can be a set of electrical impulses on a video tube. Maps appear as searchable images on touch screen displays in conference centers, airports, grocery stores, and hotels, as displays on the dashboards of cars, and as components of computer software, such as games and atlases available from the local computer store. We can think of a map independently of the technology or device by which the actual image is to be prepared. Moellering has termed such maps "virtual maps" (Moellering, 1983).

Maps of immediate interest, such as satellite and weather maps, and those showing the locations of accidents and places in the news, travel patterns, and road blockages, can be interactively received, having been updated on an hourly basis. These maps never exist other than as sets of electrons falling on phosphors, and we have had to broaden our definition of what a map is to include these new media. Probably the most common map that the average person sees on a daily basis is the weather map on TV news programs. These maps are digitally produced, with the weather forecaster standing in front of a blank wall and the camera zooming onto a computer screen showing the weather map. In addition, the actual printing of maps back onto paper is now possible using entirely new methods and materials, such as ink-jet, laser-jet, and dye-sublimation technology, and direct plotting onto film and microfilm.

1.5 NEW CARTOGRAPHIC PROBLEMS

We are moving toward automating even the toughest cartographic problems, among them the automated placement of text, an often ignored but essential part of maps and we are also increasingly making maps directly from images, especially air photos and satellite images. Dealing with text has been a particularly difficult issue for computer cartography, especially because the production of text in traditional stick-up methods was undergoing its own technological revolution due to desktop publishing, phototypesetting, laser printing, and new drafting and reproduction materials.

A map is not just a collection of lines, colors, and polygons, it also has important textual information. The selection, placement, and production of text is a very important part of cartography. Previously, this aspect was virtually ignored by almost all computer mapping systems. At last we are dealing not only with some of the issues related to putting text on maps, but we are also working to make the computer automatically select and place text where appropriate.

To be able to have the computer decide where to put the text could automate the single most time-consuming task in manual cartography. Text positioning must be such that it does not overlap other labels or important information, and that the laws of text placement are obeyed. A pioneer system for automated text placement was the AUTONAP system (Ahn, 1984) (Figure 1.1).

Once the text has been selected and placed, symbolization is critical to the esthetics of the map. Choice for the text of font, color, spacing, path and slope are variables with which individual cartographers have been able to give a map a certain "style." Much of the "artificial" look of computer-produced maps can be traced to text design. An effective mapping system would give the cartographer control over this intangible quality of maps.

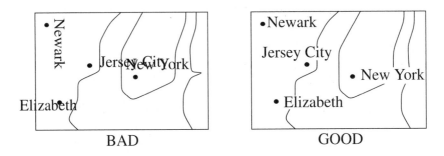

1. A name must lie totally within the bounds of the map.
2. A name should not overlap with any other name or a feature.
3. If a name overlaps with a line feature, the line, not the name, should
   be interrupted.
4. If a name is to be horizontal, it should be parallel to the rectangular
   coordinates on a large scale map.

Figure 1.1 Sample text placement rules from the AUTONAP system.

Another problem is making maps directly from images. In the past we used to go out and survey areas using plane tables. Later this was replaced with photogrammetry, using air photos to compile and update maps. Increasingly, satellites are reaching higher and higher resolutions, making them suitable for mapping at larger and larger scales, certainly scales on the order of 1:50,000.

How to make maps directly from these images without the intervention of a human interpreter is the final part of this problem. Progress is being made in this area, especially in the fields of remote sensing and image processing. Features such as roads and rivers can now be identified from imagery and replaced directly with the appropriate cartographic symbols. Mapping some of these things requires the use of artificial intelligence methods.

At the extreme, one can imagine a fully automated mapping system, acquiring data by remote sensing, extracting, identifying, symbolizing the various features in the image by understanding their context and form, and labelling the features by searching data bases by location. Some of the more recent systems show promise of these sorts of capability, providing the conceptual problems can be solved in the future.

Developments in other parallel fields are also influencing current events in computer cartography. Surveying has undergone its own computer revolution, and many surveyors now use COGO (coordinate geometry) systems, automatic note takers, and pen plotters. Similarly, the computer-aided design (CAD) industry is producing highly sophisticated tools for automating the traditional drafting process, in much the same way that paint programs are automating the artistic painting process. These technologies are influencing and being influenced by developments in computer cartography.

1.6 THE COMPUTER'S INFLUENCE ON CARTOGRAPHY

What have been the implications of these technological changes, and what has happened to mapping? First of all, cartography is now fast, or at least faster. The compilation time for a 7.5-minute quadrangle, with field surveys and checking in the 1940s, when most of the series was made, was on the order of five years. To update the maps now, we use photogrammetric methods, and the survey and field work plus the cartography takes about two years. As we move to fully digital maps, we obviously decrease the time spent in making and remaking the map.

At the time extreme, consider instead the daily weather maps distributed by NOAA over the computer networks (see Chapter 6). Every hour, a new image of North America, admittedly at low resolution but showing up-to-date atmospheric conditions, is posted on the Internet. Weather stations, television weather forecasters, pilots, and even recreational boaters can download the image and have access to cartographic data within just minutes of the image having been recorded by the satellite. Mapping speed has been increased substantially by computer mapping systems.

The computer has also increased the potential map accuracy, although accuracy is more difficult to discuss, because most people confuse *accuracy* and *precision*. If someone were to ask me the time, and I replied that it was 4:12:15.783, I would be very precise. I might be completely wrong—the time could actually be 5:31—but I would be very precise. If I were to say about half past five, then I would be fairly accurate, although imprecise.

Has computer mapping made maps more accurate? It has certainly made them more precise. We are able to store and use many more numbers now and can compute lengths and areas with more significant digits than before. A surveyor can calculate the area of a new subdivision plot to the thousandth of a hectare. Are these numbers more accurate? It certainly seems that we have taken the old causes for inaccuracy, smudgy lines and shaky hands, and replaced them with other things, such as digitizer's neck and grid interpolation error. I would argue that perhaps, in the very long term, computers have made maps more accurate. More important, however, is that our accuracy levels are now measurable against the truth, or at least against other maps, and we have therefore made our maps more accountable. Much recent cartographic research has focused on the accuracy of digital maps, and some important findings have already found their way into cartographic theory.

A major innovation in surveying is the implementation of a system involving 21 geostationary satellites in orbit around Earth, whose locations are very precisely and accurately known, because the orbits are very predictable. The system is called the global positioning system (GPS). This means that if you have a receiver that receives and decodes the signals coming from these satellites, you can fix your location very precisely by three-dimensional triangulation from a set of three satellites. It is not unknown now for a system with some microcomputer postprocessing to give your latitude, longitude, and elevation to within about 10 millimeters. That is so accurate and precise that it is good enough to measure continental drift; in fact, this technology has contributed a great deal toward figuring out that the continents are moving around and by how much. The contributions of this satellite system to the mapping sciences of geodesy and surveying are already apparent, and measurement from GPS can now be sent directly to mapping software.

This new technology allows us to resurvey things with a new level of accuracy, and sometimes we have found that we have gotten things wrong, in some cases really wrong. A good example would be the Pacific islands, totally out of sight of every other piece of land in all directions, so that locations could only be fixed by astronomical measurements of latitude and longitude. With a GPS receiver we have found out that our maps show the islands in some cases tens of kilometers off their correct locations. In many senses, cartography has a long history of islands that appear and reappear, and move around with changes in mapping technology rather than having anything to do with Atlantis, the Bermuda Triangle, or plate tectonics. A whole new promontory was discovered in Antarctica, the Washington Monument was found to have been mislocated by several meters, and probably the most famous instance is that Mount Everest spent a short period between measurements as the second-highest rather than the highest mountain on Earth.

The digital computer allows us to gather cartographic data that is both precise and accurate. Unfortunately, the implication of quality that comes with accuracy and precision has not always been justified. When the source material for a digital map is a sheet map, the digital map can become a faithful exact reproduction of all the errors involved in putting the original map onto paper. It is important to retain a "fitness for use" criterion when considering the accuracy of digital cartographic data.

1.7 BENEFITS OF THE COMPUTER TO CARTOGRAPHY

What are some of the benefits of using computers in cartography, and has the computer made mapping more cost-effective? It is rather expensive to produce maps, and in the initial stages it was believed that computers would make it incredibly inexpensive to produce maps. In fact, in the long run, computer-produced maps seem to cost about the same, with a few exceptions, as hand made maps. On the other hand, it probably will not remain that way. If we think about the stages of adopting a new technology, much of the cost of computer cartography is due to replication of the previous technology. People still use figures such as the cost of digitizing a sheet map for a single mapping project, normally a one-time fixed cost. Far lower variable costs per map are possible by reusing the digital map base.

The level of output of maps has increased using computers. Most of the fixed costs of computer cartography are in preparing a new base map, and after this step it is both easier and less costly to produce maps from the base. For example, we may spend a great deal of money producing a new census tract base map after every decennial census, but in doing so we are ensuring that we can map on that base any of the variables from the census and that we can use the map base for numerous other derived maps and analyses. Only the data, the symbolization, and map types change.

This implies that we can increase output. What this means is that mapping projects that were previously not possible are now cost-effective, even inexpensive, perhaps affordable by overseas countries and local governments, which were previously excluded by cost from the mapping business. These completely new applications have the greatest potential for computer cartography, for these users have no previous technology to waste time replicating and they can reap the benefits of other peoples' experience.

The overall effects of the computer on the discipline of cartography are many. Cartography in the 1950s was much of a service discipline. A cartographer would be attached to a geography or geology program academically, or isolated in a map department commercially. A cartographer's function was as a technician, a producer of maps, rather than as a producer of systems for making maps. Wolter (1975) argued in *The Emerging Discipline of Cartography* that cartography has evolved, through its self-examination as a result of the computer age, all the requirements to become a distinct academic discipline in its own right. Today, the distinctiveness of the discipline and its new role are clear.

Certainly, if one examines the literature, this is evident. Much that is published in cartographic journals these days is clearly analytical cartography, although there are other approaches to the discipline (Moellering, 1991). One could argue that this is just a return to the preeminence of cartography, a return to the role that cartography held under the ancient Greeks or during the age of discovery. Many more people outside the discipline now encounter and eventually study cartography. As a result of this scrutiny from outside, we have been forced to define the cartographer's domain of interest very precisely. We have had to define in detail exactly what cartographers do, how they do it, exactly what the information they deal with is, and what its limitations are. And we have had to model the flow of this information through the mapping process.

The computer has relieved the cartographer from tedious production tasks. Anyone who has experience with manual cartography knows how much hard work goes into a manually produced map, even a poorly designed or inexpertly drafted one. The focus of cartography used to be on the production technology. If the computer is doing the minor production tasks for us, we can worry about other things, such as design—which is ultimately what a cartographer is trying to perfect. For one thing, the computer has allowed us to set up a design loop. The manual equivalent is to work on a series of separations one at a time, incorporating a single map design for each map project. We finally get to producing a map proof, and for the first time after perhaps 100 hours of work, get to see what the map looks like. What if we don't like what we see? More commonly, we might say, "If only this text were slightly smaller" or, "If only this red were green." It is too late to change the map; in other words, we have no design loop.

In a design loop, the finished product can be interactively modified to make all the changes we may wish to incorporate. In fact, we can use the design loop to produce experimental maps that will help us to learn better design: for example, to see why a map with 12 fonts and 30 shade classes really does look bad. This changes the emphasis from the technology to the design and further to the principles that make this a good design regardless of the technology. In this way, we can make maps look better — which is not a bad definition of what it is that cartographers now do.

The computer has produced a whole series of new capabilities. The ability to manipulate color is a good example of this. The computer has given us new types of maps and new media for their display. The field of scientific visualization has produced maps that can be walked around, manipulated and perceived as three-dimensional objects. Cartographers are now beginning to exploit the tools and methods of the new interactive media, multimedia, and animation to show spatial distributions over time and space (Andrews and Tilton, 1993). Similarly, we can use cartographic techniques to map new types of data beyond the traditional domain of cartography, such as statistical distributions, the ocean floor, Mars, or a molecule (Hall, 1992).

1.8 DISADVANTAGES OF THE COMPUTER FOR CARTOGRAPHY

The computer has had some negative effects on cartography. For one thing, the amount of technical training that a cartographer has to acquire has increased enormously. The cartographer of the 1990s must be a database expert, a user-interface designer, and a software engineer. He or she must also retain a sense of map esthetics, be familiar with diverse computing environments, and still produce maps. Cartographers ideally need training in image processing, remote sensing, photogrammetry, land information systems, geographic information systems, surveying and geodesy, and computer programming.

Another disadvantage of the computer for cartography as a discipline is that those not trained in cartography can now easily produce maps. Although in many cases this is a popularization of the mapping process, in the early days of computer cartography it led to a renewed need for good design and esthetics. For a time, in replicating the previous technology, mapmakers threw out the quality standards that the previous technology had contributed to cartography as a whole. For about 20 years computer cartography produced rather inferior products, which were accepted not because they were better but because they were new and different. Today, digitally produced maps are usually superior esthetically to those we can make by hand.

1.9 SOME GENERAL TRENDS

There have been many general trends within cartography as a result of the advent of the computer. One trend has been toward simpler maps. If we were stuck with a single product, we often believed that there were savings to be made by incorporating as much data as possible on a single map. Maps have become "single message," in much the same way as business graphics, with a good example being the weather map or the sales map. These are maps being used to communicate a message. A whole school of thought within cartography has concentrated on the effectiveness of maps in communicating a particular message to the map reader. This communication school has sought to improve map design to get the message across faster and more effectively.

The computer has given us defensible design. We can assign numbers to back up qualitative judgments. A good example is to use recognized error reduction techniques to choose class breaks for shaded maps or known methods to do line generalization and simplification. So an island may be eliminated on a map at a certain scale because it falls below a certain threshold area, rather than because it was easy to paint over the island with opaquing fluid on a negative.

The computer has promoted new types of cartographic symbolization and representation; that is, we have devised new types of maps as well as new ways of making the old types. An example is the depiction of terrain. Now we can generate realistic depictions of terrain to simulate the way the terrain will actually look to an observer. In particular, the computer has made thematic mapping available to those who are not cartographers but who work with statistical data with a geographic component. The computer, therefore, has opened up mapping to those without a formal training in the discipline. This may yet prove to be the most critical advance in the acceptance of computer mapping.

Finally, education has expanded cartography to include computer and analytical aspects. We have had an enormous increase in both the numbers of, and demand for, the type of college course at which this book is aimed. Almost every geography program in the country offers at least some form of computer (not necessarily analytical) cartography. Sometimes schools offer multisemester sequences, even entire specializations in computer cartography, usually coupled with sequences in geographic information systems and remote sensing. Increasingly, the graduates of these classes go on to find employment in the booming industry of computer cartography, well endowed, one hopes, with a sound base in the analytical aspects of the discipline, which will ensure those same students of employment in the future.

1.10 GEOGRAPHIC INFORMATION SYSTEMS

The influence of the computer on cartography has opened up cartographic methods and techniques to more general purpose uses of geographic information. Nowhere has this been more evident than in the field of geographic information systems (GIS). GISs are automated systems for the capture, storage, retrieval, analysis, and display of spatial data. Large numbers of these systems are now available to assist resource and spatial data managers with their planning and decision making, as well as their routine record keeping and inventory. Because these systems use geographic data extensively, most of the queries made of these systems provide cartographic solutions.

As a result, these systems are usually supplied with either an interface to a separate cartographic display system or contain their own such systems. The type, quality, and capabilities of these display systems vary remarkably from system to system, ranging from poor to the very best in quality. The major differences in how GISs generate maps are twofold. First, the map is either an intermediate or partial solution to a query and may therefore not contain any of the true cartographic elements other than a figure and a georeference system. This is not to say that the final maps that provide the problem solutions are not good cartography, but these maps instead reflect the emphasis on query and response rather than display.

Second, the user of a GIS is not always a cartographer and therefore does not place the same level of importance on the map as on the message the map conveys. GISs have become the primary way in which people are first introduced to computer cartography, and in some cases to analytical cartography. It is important to realize, however, that the goals of computer cartography and the goals of geographic information processing are not always identical, although they are always similar.

The skilled cartographer should develop an understanding of GISs and should see in them the challenge of a new set of demands on the theory and methods of cartography. It is not sufficient that the computer cartographer is interested solely in the graphic display of digital cartographic data, since the structuring, encoding, and the representation of the

data are the very basis upon which analytical cartography is built. Cartographic transformations, the basis of analytical cartography, underlie all geographic data processing and
therefore are of overlapping interest to cartographers and the builders and users of GISs
alike.

1.11 REFERENCES

Ahn, J. K. (1984). *Automatic Name Placement System*. Publication No. IPL-TR-063, Image Processing Laboratory, Rensselaer Polytechnic Institute, Troy, NY.

Andrews, S. K., and D. W. Tilton, (1993). "How Multimedia and Hypermedia Are
Changing the Look of Maps." *Proceedings, AUTOCARTO 11, Eleventh International Symposium on Computer Assisted Cartography,* Minneapolis, pp.
348–366.

Hall, S. S. (1992). *Mapping the Next Millennium: The Discovery of New Geographies.*
New York: Random House.

Moellering, H. (1983). "Designing Interactive Cartographic Systems Using the Concepts
of Real and Virtual Maps." *Proceedings, AUTOCARTO 6, Sixth International
Symposium on Computer-Assisted Cartography,* Ottawa, October 16–21, vol.
2, pp. 53–64.

Moellering, H., ed. (1991). "Special Content: Analytical Cartography." *Cartography and
Geographic Information Systems,* vol. 18, no. 1.

Morrison, J. L. (1980). "Computer Technology and Cartographic Change." in *The Computer in Contemporary Cartography,* edited by D. R. F. Taylor. New York:
Wiley: Chapter 2, pp. 5–23.

Morrison, J. L. (1993). "Cartography and the Spatially Literate Populace of the 21st.
Century," *Cartography and Geographic Information Systems,* vol. 20, no. 4,
pp. 204–209.

Tobler, W. R. (1959). "Automation and Cartography." *Geographical Review,* vol. 49,
pp. 526–534.

Tobler, W. R. (1976). "Analytical Cartography." *American Cartographer,* vol. 3, no. 1,
pp. 21–31.

Wolter, J. A. (1975). *The Emerging Discipline of Cartography.* Department of Geography, University of Minnesota, Ph.D. Dissertation, University Microfilms, Ann
Arbor, MI.

2

Hardware

2.1 DEVICES FOR COMPUTER CARTOGRAPHY

2.1.1 A Map of the Computer

In Chapter 1 we saw that the impact of the digital computer upon cartography has been revolutionary. To understand the tools of contemporary cartography, therefore, we must examine computers in depth. We shall begin by considering the nut and bolts of computing, the hardware. This will allow us to focus on the special types of devices required for computer cartography, most often unlike those required for more routine computer tasks. To conclude the chapter, the relatively recent addition of the workstation will be introduced, and its significance for both analytical and computer cartography demonstrated.

Hardware consists of a set of physical devices that perform electrical, mechanical, or other physical tasks, or state changes according to instructions given by software. Hardware is something breakable, needing mechanical assistance, repair, or maintenance, and it reflects a particular stage of technology, therefore implying eventual obsolescence. As a result, the center of a computer is of relatively minor importance for analytical cartography, although of some importance for computer cartography. Of more importance to cartography is the peripheral device, a plug-in or add-on part of the computer that gives the computer cartographic capabilities. A recent trend has seen peripheral devices with significant amounts of computing power, sometimes more than the "main" computer itself.

Figure 2.1 presents a *map of the computer*. At the heart of the computer is the central processing unit (CPU). The CPU consists of the part of the computer that actually carries out the instructions given by software, such as the addition of values in registers, places where binary operations take place. In a microcomputer, the CPU can be as small as a

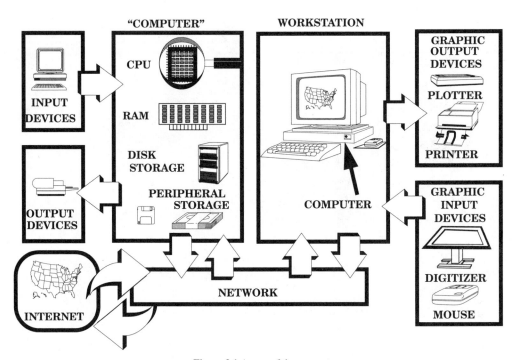

Figure 2.1 A map of the computer.

single microcircuit or "chip." CPUs vary in power according to the size of their registers in bits and their speeds, usually measured by their clock speeds in megahertz, their basic processing rate in MIPS (millions of instructions per second), and by their ability to handle complex operations (megaflops, or millions of floating point operations per second). Many CPUs are enhanced with an accelerator, which in particular increases the speed at which the CPU can handle floating point arithmetic. Most computers can do integer arithmetic at high speeds, but slow down considerably when the use of floating point causes the registers to overflow.

The memory locations accessed by the CPU have to match the speed at which the CPU operates and therefore are limited in size. In addition, this memory must be able to store instructions, data, and parts of the operating system. Such "fast" memory must be accessible starting at any point and is therefore known as random access memory (RAM). While the first microcomputers were limited by the size of the link to the CPU (the bus) to about 28 kilobytes or less, most microcomputers now have over 640 kilobytes, with some having well over a megabyte. Workstations typically have about 8 to 32 megabytes of RAM. Because the size of the RAM is the upper limit on the address space available to computer programs, this is often a limiting factor for cartographic software.

Data and instructions awaiting the services of the RAM and the CPU reside in one of the various forms of permanent or temporary storage. In large computers, many different kinds of storage are available. Among the more temporary storage mechanisms are fixed and removable disks. Fixed disks are installed within the computer and involve a mechanical device capable of writing information onto and retrieving it from a storage medium. Magnetic platters, optical disks, and even "bubble" memory are used to store information. Removable systems allow data and programs to be archived or added to the system. Removable systems include disk cartridges, magnetic tapes, floppy disks, and removable optical media. Most cartographic software and data are available on microdisk or magnetic tape. Among the disks, the 133 millimeter (5.25 inch) and the 90 millimeter (3.5 inch), are standard. Among magnetic tapes, the nine-track 13 millimeter (0.5 inch) at the 1600- and 6250-bit per inch densities are most common, while workstations usually use cartridge tapes. Mass storage is usually by 8-mm tape, an extremely inexpensive storage medium onto which about 5 gigabytes per tape can be written. For read-only purposes, the CD-ROM is the preferred medium, because it is robust, inexpensive, and easily produced and distributed. These media are comparatively cheap, widely available, and reliable enough to store data without error for several years.

Computer cartography involves interaction with the computer in ways that are more complex than the simple video terminal and keyboard, which microcomputers have made household features. In particular, we need to enter graphic locational information into the computer and to get output in the form of maps from the computer. This involves adding to the computer devices that are specially made for graphics. Such add-on features are called peripherals, although this term includes add-on memory, disks, keyboards, mice, and so forth. A more specific term for a peripheral is a *device,* and the devices we require for cartography are graphic input and output devices. This chapter is, therefore, largely a summary of the various graphic input and output devices that have been used for computer cartography. Historically, these devices were used in isolation, connected remotely to a large computer, and performed their functions one at a time, so that their use in cartography was much the same as when maps were produced manually. More recent developments, however, have reshaped the way devices are used. This development, in turn, has been made possible by two other developments, the development of computer networks and the development of the workstation.

Computer *networks* are groups of linked computers that can share the various devices that are connected to the network rather than to individual computers. Within a network, each CPU, with its RAM, becomes a client and is connected to the network via a cable and network software, which handles the movement of data through the network and manages the individual systems. Peripheral devices (such as disk storage, tape drives, and especially input and output devices) can then be connected to the network rather than simply to an isolated CPU. One of the CPUs on the network typically functions as the master, usually supplying most of the disk storage and services. This CPU is known as the host, and if it supplies the disk space to the network, it is known as the file server. In the case of computer cartographic software, installation is often on the file server. The software is then "released" to all of the users of the network.

Networks can be extended beyond the limits of one system by telephone and other links to other computers. Thus entire databases can be distributed, resident on different computers around the United States, or even the world, and used by clients remotely. Such networks are available even to the smallest system, including microcomputers, assuming the availability of a modem.

The second important development for cartography is the graphic workstation. In the most primitive form, a system to produce graphics contains only additional RAM connected to a graphic display card, and a monitor capable of supporting graphic display. A workstation, however, can support any number of graphic input and output devices simultaneously. As a minimum, the workstation should have both a graphics and a text display and some sort of graphic input device, such as a digitizer, mouse, or touchscreen. Many workstations incorporate more sophisticated devices. A workstation has its own operating environment and is usually connected to a network, where data and additional software are available.

2.1.2 Input Devices

In the preceding subsection we met the concept of a peripheral device. In computer cartography, the devices of major concern can be classified as input, those that provide data, and output, those that produce maps or other information. Within this division, we can break the devices down further into map and attribute.

Attribute input devices, although in regular use, are not particularly cartographic. Examples are the keyboard, a computer tape, or a disk from which a file can be read. Anything that can send data to a computer qualifies as an input device. Map input devices are many and varied. The most primitive, and first, was the up, down, left, and right arrow keys on a keyboard, sometimes mapped onto the h,j,k, and l keys. Continuing chronologically, the next was the joystick. Joysticks were used occasionally in computer cartography, but remain in use solely for games. Some graphics terminals included on the keyboards two wheels that moved a set of cross hairs on the screen in the x and y directions. Positioning the cross hairs and then touching a key returned the screen location to the device. This type of input has largely been replaced with the light pen, digitizing pad, touch screen, track ball, and mouse.

The light pen is favored in CAD applications, where interactive screen modification of a picture is important. The light pen is much like a pen, except that the point end emits a light signal that is detected by the screen and the other trails a wire. The digitizing pad is a small digitizing tablet, sometimes only about 200 millimeters on a side. Usually, the tablet includes, or has attached to it, a menu from which commands can be selected by pointing to the menu area indicted and either pressing a button on a cursor, or in the case of a pen cursor, pushing down on the pen tip. The graphic interaction takes place away from the screen, although when interactive graphics use the screen, the cursor position is usually continuously displayed on the screen. Many "paint" packages, the digital equivalent of a pen and paintbrush, use the digitizer pad. Figure 2.2 shows a workstation with a full set of input devices.

Another form of graphic interaction is the touch screen. On a touch screen, the screen is partitioned into menu areas, on which is displayed text or graphics describing the selection. When the user touches the screen, the touch is detected and the software reacts accordingly. Touch screens have found their way into supermarket guides, hotel and airport lobbies, and libraries, where searches are a set of hierarchically linked choices, narrowing in on a final image to be displayed. In many cases, these systems include maps.

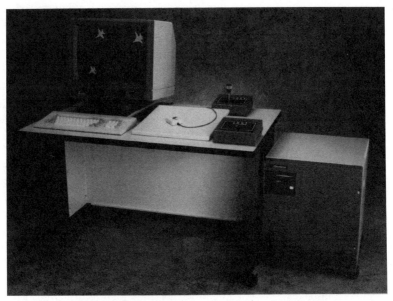

Figure 2.2 Workstation with multiple input capabilities.
(Photo permission of Chromatics, Inc.)

For example, a hotel lobby screen may contain a neighborhood map with restaurants, which can be queried by the ethnicity of the food served.

The track ball is a ball mounted in a holder, which can be moved freely in any direction using the palm of the hand. Often buttons or dials allow the motion to be constrained to one direction or to proceed just one screen unit at a time. As the ball is moved, the screen cursor moves with the same relative movements as the ball. Most track balls use "wraparound"; that is, when the cursor is moved off the screen to the edge, it reappears at the oppostite side.

By far the most common graphics input device is the mouse. Two types are in use: the mechanical and the optical. The mechanical mouse contains a small plastic or rubber ball that is turned by friction against the table top or a special mouse pad as the mouse is moved. The relative movements are picked up by the computer and are used to control the motion of a cursor on the screen. The optical mouse uses a reflective pad and a light beam to detect motion, which avoids the problem of dirt and dust clogging the ball as sometimes occurs with a mechanical mouse. Cordless mice allow complete freedom of movement since they need no wire connection to the keyboard.

Using a mouse involves first moving the cursor around, but also selecting features from a graphic menu. Items are selected by clicking, that is, moving the cursor into the correct location and pressing a button on the mouse. Often, double-clicking (making two clicks of the mouse button in rapid succession) makes the selection and confirms it. Holding down a button sometimes reveals another level of menu, the so-called pull-down or pull-right menus used in many graphics-based operating systems, which require more cursor movements. Selecting a feature by holding down a button while the cursor is over a feature and then moving the feature by moving the mouse while holding down the selection button is called dragging. Dragging is an integral part of using CAD and draw-type software.

What the map input devices covered so far have in common is that although they have some graphic capabilities, the locational data they provide is too approximate for anything other than selection from menus. These devices support the graphics function of *selection,* that is, indicating approximate direction and location, and of *picking,* the choosing of a menu object. Detailed cartographic input comes from one of two sources, semiautomatic digitizing and automatic digitizing.

Semiautomatic digitizers are the computer equivalent of the drafting table. They conaint a tablet, that can be as small as a printed page, or as large as 1.5 by 2 meters. Smaller tablets can rest on a desk top, but the larger ones require stands and include options such as power movement and back lighting. Although a number of different options are available for the cursor on a digitizer, most cursors have between 1 and 16 buttons and have options such as magnifying lenses, choices of bull's-eye and cross-hair patterns, and the ability to follow lines within the cursor window automatically. Figure 2.3 shows a typical array of tablet and cursor choices. Cursors with multiple buttons allow data such as elevations and labels to be entered along with the map information, which means that the user need not move between the keyboard and the tablet.

Generally, the tablet itself consists of wires embedded in the surface along the cartesian axes. Each wire is connected along the edges, so that when the cursor is moved to a point and the button is pressed, an electrical charge is generated by a coil in the cursor which is detected on one of the wires underneath. The unique wire that picked up the charge gives the x and y values. The signal is then converted to a string of ASCII characters corresponding to the button pressed, the x and y values of the point, and any attributes entered. These data are sent down a cable to the controlling computer, usually a workstation or a microcomputer.

Early systems often converted the signal to codes for transmission over telephone lines, because much early digitizing took place with remote connections to large computers or directly to magnetic tape off-line. This arrangement has been replaced so that the signal from the tablet now moves straight to a special purpose board inside a microcomputer or the worstation that controls the tablet. In this case, the data move directly to disk.

Electronics is not the only technology that has been used in digitizing. For a long time there were sonic digitizing tablets, still seen occasionally. Sonic tablets use a drafting table, but instead of wires set into the tablet there is a string of microphones along the top and one side. The button on the cursor simply emits a click or a beep, which is picked up by the microphones. A triangulation is then performed to figure out where the cursor is. Other tablets are simply mechanical arms like pantographs, which use the mechanical movement of the arms to compute x and y. These systems are usually inexpensive and can be mounted on an existing table.

Many digitizing tablets support voice data input in addition to the cursor. Simple commands can be spoken, and a microphone on the controlling computer matches the recorded voice against a preset list of commands. This allows much more efficient use of the tablet because the user's attention remains on the task, not on moving between menus and

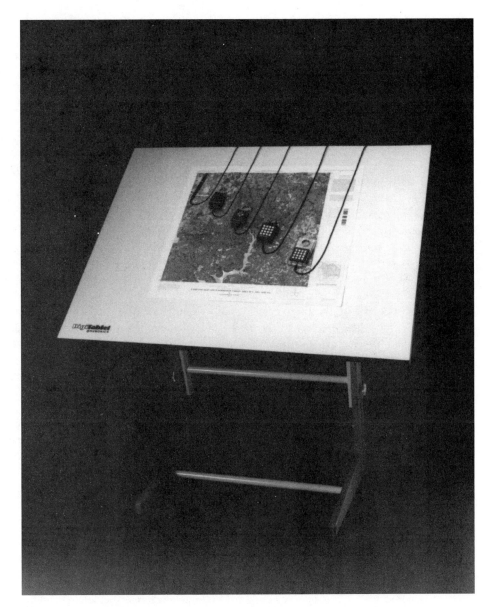

Figure 2.3 Digitizing tablet. (Photo courtesy of Numonics.)

cursor buttons. Also in use are three-dimensional digitizers, where a cursor on a cantilever is moved around an object touching the edges to record three locational dimensions. These systems have been little used in cartography.

Semiautomated digitizers are most commonly used when maps are concerned, and the larger sizes are required for digitizing sheet maps. The alternative technology, *automatic digitizing,* is more recent. Automatic digitizers include scanners and line followers. At the low-cost end of the scanner market, two types of scanner exist. The first, which were designed to scan pages of text and automatically convert the pages to text files in the computer, are not suitable for cartography, with the possible exception of producing text lists for typesetting. Some of these scanners are capable of low-resolution graphics, and parts of small maps can be scanned into paint and draw software using this technology. Most of these scanners operate in pixel mode in monochrome, but some conversion to vector or object mode is now possible.

The alternative scanning technology is the video scanner. The video scanner is the digital equivalent of a television camera and is mounted on a stand. Maps, images, or air photos can be placed under the camera, and the television picture is shown on a monitor and rasterized, that is, it is turned into a grid of numbers. An identical technology exists for still cameras, which store their images on small disks and replay them through a video player as television images. Normally, what is returned is exactly what comes back from a remote sensing device, the gray tone or the intensity in a certain light frequency (a certain color) for each little square area or pixel on the image. Red, green, and blue channels for a color map can be scanned separately and re-superimposed to generate color from three gray-scale images.

Geocoding on these images is usually established using reference or control points and rubber sheeting, topics covered in later chapters. Only small areas of maps can be scanned with these low-cost systems with true cartographic fidelity, although the resolution capabilities of these systems are improving rapidly.

The top-end scanners are of very high resolution and as a result are expensive, although their price has fallen considerably recently. High-precision scanning devices use what is called a flying spot technique, which involves moving a light or laser beam very quickly across the surface and detecting either absorption or reflectance at every point on the surface. Precision scanners often rotate the map on a drum, moving the laser in very small increments across the map surface. Because paper maps can fold, fade, or tear, the source map is usually the film separation from which the map was printed. Monochrome scanners can be used to scan each ink-color separation separately for any map. Increasingly, however, color scanners are capable of extracting colored features directly.

One of the problems with scanning maps is that most maps have a great deal of redundancy. Polygon maps contain information only at the boundaries. Also, text is not easily processed from scanned data. Other problems mean that much postprocessing must be conducted on scanned maps. For example, the map separation containing contour lines can be scanned simply, yet the breaks in the lines for contour labels must be joined together after scanning to give continuity and the labels must be deleted and turned into an accompanying attribute file. Also, scanners often have resolutions that supply an immense amount of data for very small pixels. As seen in Part II, geocoding maps involves far more than simply converting a map into sets of numbers.

Scanners are in widespread use for computer cartography. Many of the earlier problems with scanning have been eliminated. An important breakthrough was the development of hardware for the automatic conversion of the scanned data to strings of (x, y) pairs, known as vector data. These special purpose vectorizers have made scanning a valid alternative for digitizing in bulk. Scanning is widely used in high-quality production cartography. As the costs and capabilities of scanning evolve, it is highly likely that scanners will assume most, if not all, of the map input process in computer cartography.

2.1.3 Output Devices

As before, we can divide output devices into map and attribute. Good examples of the latter are such things as the line printer, disk, tape, and any other nongraphic medium. The line printer was used as an output device by the very first generation of automated mapping systems. The limited capabilities of the line printer were augmented by overstriking characters to get gray tones, by printing symbols, and by retouching the final output.

Contemporary map output devices fall into several types. The first type are plotters. Plotters are an example of how computer cartography tried to replicate the previous technology. Plotters do this by drawing a map on paper. If digitizing can be thought of as an analog-to-digital conversion, then plotters convert the digital back to analog information. Analog (especially paper) maps have some advantages for storage and are convenient to carry, especially into the field.

Pen plotters fall into two different types, flatbed and drum. For many years flatbed plotters were cheap, and readily available but not really of high cartographic quality. Drum plotters were expensive, and difficult to operate, but produced much higher precision, better-quality output. Now, however, some extremely inexpensive drum plotters are available, while some of the very highest quality cartographic output is now plotted, or more usually scribed, by flatbed plotters. Both flatbed and drum plotters are vector devices. Flatbed plotters take a sheet of paper, film, or scribecoat and attach it to a flat surface. The pen then moves over the paper, which is held in place with tape or more frequently by an electrostatic field that holds the paper perfectly flat. The plotter then simply moves a pen along two axes.

Different technologies determine the actual drawing. The pens can move on an arm, which itself moves along an axis; the "bed" can move in one direction and the pen in the other; or any combination of these movements is possible. All plotters have only two fundamental basic operations: move with the pen down or move with the pen up. If you move around with the pen down, it draws.

With a flatbed plotter you typically have to load each sheet of paper or film yourself and take it out again when it is finished. You also have to load the pens yourself, although even the least expensive plotters are capable of changing pens automatically. More sophisticated plotters can recap pens and change ink without operator assistance. Also available are automatic paper feed and the cutting of plots when the drawing is complete. This allows the use of rolled paper and the queuing of multiple plots for plotting when the operator is absent.

The drum plotter (Figure 2.4) has a rotating drum with prongs that feed holed paper from a roll. Drum plotters have hoppers that hold one or more pens to support different colors. Motion in the x direction is achieved by moving the paper with the drum past the pens. Motion in the y direction is by moving the hopper across the drum (Figure 2.5). Most drum plotters can use a large variety of pens and plot media. Pencil point, fiber

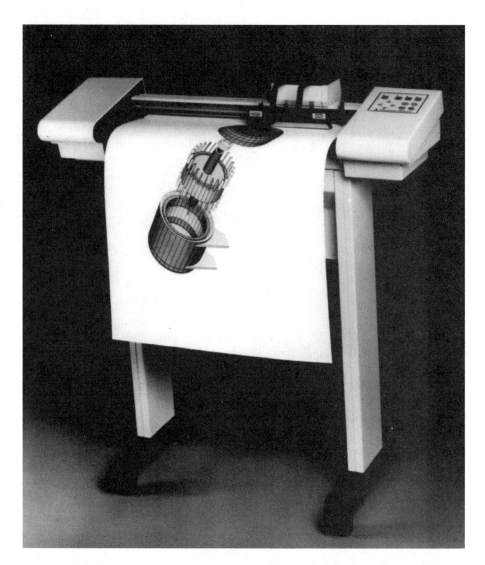

Figure 2.4 A drum plotter. (Photo permission of Houston Instruments, Austin, TX.)

point, rollerball, and refillable ink pens can be used to plot onto paper, film, Mylar, or acetate. Some very high quality maps can result. Both flatbed and drum plotters, however, are poor at filling areas on maps with continuous color, because the pen has to be dragged backwards and forwards until the area is filled. When all that is required is text and linework, the quality is of cartographic standard.

Another type of plotter is the electrostatic plotter. Probably the major uses of electrostatic plotters are in the electronics industry, where they are used for printing multicolor circuit boards because they can print a large area of coverage with a great deal of detail.

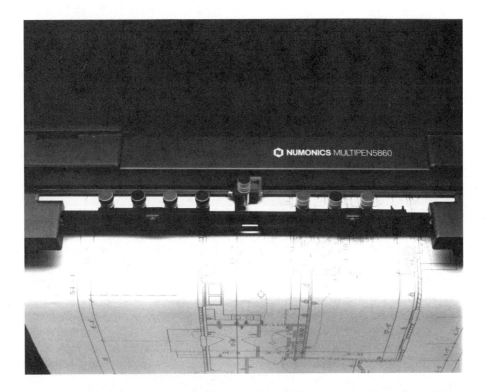

Figure 2.5 Pen hopper on a drum plotter. (By permission of Numonics Corp., Montgomery, PA.)

Electrostatic plotters use electrical charges to attract ink to the surface of treated paper. Although electrostatic plotters are bulky, expensive, difficult to maintain, and often support only a few colors, they are better than pen plotters at areal coverage.

Moderately related to the electrostatic plotters are printing technologies. Printers use impact through an ink source onto the surface to produce a plot as in a typewriter. The first printers of near-cartographic quality were dot matrix printers. More recently, laser-jet printers have taken over completely. Laser-jet printers have high definition, usually as simply black on white, and have revolutionized desktop publishing and word processing. The resolutions of low-end laser-jets is several hundred dots per inch, but these printers are fully consistent with typesetting-quality printers with over 1,000 dots per inch.

Many pieces of cartographic software use the laser-jet printer as the primary, or even the only, graphic output device. The PostScript industry standard, widely used by laser-jet printers, has allowed a great deal of device independence, and many drivers for graphics allow maps to be sent to laser-jet printers. The laser-jet printer is a refinement of the xerographic process. The link between publishing, cartography, and typography that this technology offers is another opportunity for cartographers to make superior maps.

Impact printers are capable of low-quality color. Better-quality color is available, however, without photography using a hybrid technology halfway between dot matrix and laser-jet plotters. Ink-jet plotters are a hybrid because they are not printers and they are not plotters. They use a method where ink is sprayed directly onto the surface. They normally have four ink guns with four different inks for the production primary colors of magenta, cyan, and yellow. The fourth gun contains black. Mixtures of these four inks give a very large number of different color possibilities. High-end ink-jet plotters produce almost photographic quality images and are getting better all the time. Low-end ink-jet plotters, however, produce rather garish color, can print only fairly narrow strips of paper, and are of low resolution.

Two new technologies have led to significant cost reductions in high-quality printing. The first is the bubble-jet printer, a technology in which inks are sprayed onto the paper in much the same way as in ink-jet printing, but the final propulsion onto the paper is accomplished by the rapid intense heating of the ink in a narrow jet. Bubble-jet printers can produce large formats and color and are comparable in quality but are generally less expensive than laser-jet printers.

Another significant development is the development of dye-sublimation printing, in which dyes are placed onto prepared paper under natural light. Dye-sublimation printers can give extremely high-quality, photograph-like output on glossy photographic-type paper for a very reasonable price. Although these printers still cost thousands of dollars to purchase, many slide and copy services now have this technology and can produce output in this form.

The last major category of map output devices are those involving photography. At the high end of film technology is the film writer, which typically writes onto a very large piece of film and thus is favored for the hard-copy of remotely sensed images. At the middle of the range is the graphics camera. Graphics cameras are separate devices connected to the red, green, and blue color channels normally sent from a graphics system to a color monitor. The color image is then reconstructed on a flat display at the end of a box, in the focal plane of a bolt-on camera back.

Typically, the exposure is controlled completely by a microprocessor, including the aperture, the light intensities, and the exposure time. Each of the red, green, and blue channels is exposed separately, making the color image additive. Graphics cameras usually support most photographic media, including 35-mm film, the various instant picture formats, and various sizes of sheet film. The attraction of this method of hard-copy is the price, which can be very reasonable per image, and that the images are virtually distortion free. Service bureaus now also offer extremely low cost conversion of digital files from raster formats to 35-mm slides or to photo-style prints. This is often far cheaper than purchasing even the lower-end graphics film copiers and recorders.

At the low end of the camera market are screen copiers, some of which are very inexpensive. These devices mount onto a black screened box that fits over a monitor. The mount supports usually an instant camera type of camera back, although some support 35-mm film. A number of different processes can be used, including one type of slide film that can be developed in only a few minutes with a stand-alone slide developer. The quality of this type of output is sometime very good, although care must be taken to avoid the "barrel effect" of cameras that are focused too closely onto a surface that is actually curved and not flat.

True flat screens, at first only as liquid crystal displays (LCDs) but now in color and as back-lit plasma screens, offer distinct advantages to cartography. Last but not least, an inexpensive way to photograph computer maps is to focus a 35-mm camera with a tele-photo lens and a tripod onto the screen from a distance of about 2 meters in a darkened room. An $f/4$ and about 1 second for ASA 64 film is required because the scan lines are caught in the image at speeds greater than about $\frac{1}{15}$ of a second. Any reflections or screen edges in the image can be masked out on the slide with tape.

As a final example, some plotters now plot directly onto microfiche and microfilm. This may be one of the major ways that analog maps will be distributed in the future, es-pecially if a hand-held viewer is used. The international report to the ICA, published in the *American Cartographer* for 1987, includes two microfiche maps submitted as the best examples of American cartography in the previous year. Digital maps can be produced directly from plotting devices onto microfiche or microfilm with no paper involved. These media take up little storage space, and can be mailed in a single envelope instead of a map tube, eliminating rolling and folding.

2.2 RASTER AND VECTOR DISPLAY TECHNOLOGIES

A major historical distinction between different computer devices for computer cartogra-phy is that between raster and vector technology. These two different display technolo-gies have had a particular influence on much of the theory and data structures involved in analytical and computer cartography. It is worth describing the alternative technolo-gies in a developmental context so as to introduce the newer technology, which has elim-inated many of the inconsistencies between the two technologies.

In raster technology, the screen or the input device is divided into discrete pieces called pixels, which have a finite size. For an orbiting scanner on a spacecraft these would be single areas on the earth's surface for which we can resolve data. On a map they could be latitude and longitude cells or meters marked off on a Universal Transverse Mercator (UTM) grid. Each square or rectangular pixel has an attribute associated with it, which could be a color, a light intensity, a gray tone, or a red, green, blue value for that pixel. For input they would be associated with readings or measurements. In terms of output, the data could be displayed as is, each pixel mapped onto a piece of the screen.

The alternative is vector technology, in which control of the output device comes from software, which moves a point light or pen across the screen or paper, leaving a mark. Vector input is particularly suited to maps that have lines on them, such as coast-lines or rivers, political districts, contours, and so forth. Raster input is particularly suited for maps that have attributes with extensive areal characteristics such as land use, land cover, surface moisture or ocean temperature: an attribute measured over a large area. Vector and raster, therefore, are two different structures, two different technologies for both input and output.

Vector graphics dominated the computer graphics market very early on, with storage tube-technology terminals. A storage tube contained an electron gun that sent a stream of electrons toward a screen treated with different phosphors that glow when struck with electrons. Control over movement of the beam was achieved using a series of magnets which allowed the beam to be deflected and moved continuously. Because the phosphors stored the charge when struck by electrons, the screen stayed glowing when hit, and lines, points, and areas could be traced out with the electron beam. When complete, the entire

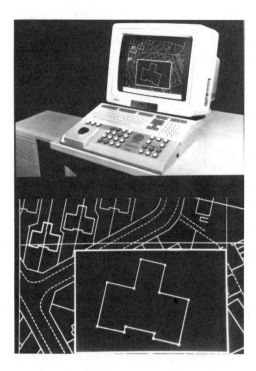

Figure 2.6 Special purpose raster-to-vector converter. (Photo courtesy of
LaserScan Laboratories Ltd., Cambridge, England.)

screen had to be cleared at once, usually by depressing a button on the keyboard. Storage tubes allowed straight and curved lines to be traced out, although they were poor at filling in areas. In addition, virtually all storage tubes were monochrome, usually a light green glow on a dark green background.

The advantage of vector displays is that we can draw a line in any direction with any length and have it appear exactly the way it would on a paper map. The finite limit on resolution is the size of the grid of phosphors, a very small spot on the screen surface. The disadvantages are several. First, the devices have to be fairly large. My first graphics terminal, a storage tube device, was so large that it had to have its own stand. The display tube on the terminal had to be quite deep, and the screen area, although flat, was very small. In addition, the phosphors eventually decayed, leaving images "burned" into the screen. Storage tubes also could be quite slow in drawing a map, although plotting time was a function of the power of the computer behind the terminal. Because selected update was not possible, waiting for redraws could be tedious. Raster technology now completely dominates the computer graphics industry to such an extent that software when required usually simulates a vector display device using a raster display. Raster technology evolved out of television technology. In this case the screen phosphors respond differentially to electrons by changing red, green, or blue in different intensities. A major difference is that the phosphors are arrayed in a grid as pixels. A look at a color television screen using a magnifying glass will reveal the cells that can illuminate in different colors. With a very large number of pixels, they appear to give a continuous image, although in fact it is a discrete image. In raster displays, the electron beam scans the screen with about one full scan every $\frac{1}{60}$ of a second, refreshing or temporarily illuminating each in turn. To actually see the screen redrawing we have to use something that is fast enough that it sees it only for a very short space of time. A good way to do that is to use your 35-mm camera and take a screen photograph. At $\frac{1}{30}$ of a second, bands are visible in screen photographs across the picture or the TV screen because while the shutter was open, the screen refreshed once and got part way through refreshing a second time before the shutter closed.

The conversion from raster data (preferable for scanning) to vector data (often preferable for plotting) is often performed by hardware rather than by software. In Chapter 11 the vector-to-raster (and vice versa) transformations are considered as a software problem. The same task, however, is commonly performed by special purpose devices. One such system is shown in Figure 2.6. Because the vectorization often introduces ambiguities, especially in the topology of line and polygon maps, these devices must support extensive operator control over editing. It seems likely that human interaction will be used in this editing and matching process for some time to come.

2.3 REAL AND VIRTUAL MAPS

With so many different pieces of hardware, capable of producing so many different types of digital map products, it is not inappropriate to ask "What is a map?" To answer part of the question, Moellering (1980) introduced the concept of real and virtual maps. A real map is one that has gone through the symbolization transformation and has found a realization as a tangible object. This implies that the cartographer selected a scale and plotted a specific map type, legend, and text. A virtual map is like a possible map; for example, if we have a digital outline map of the United States and access to state statistical data, then any variable can be mapped, using a number of different methods, classifications, and so forth. The set of possible maps is finite and even directly accessible using an interactive mapping package. A virtual map, therefore, is both a nonpermanent map in itself and also a digital equivalent of a "proof" map, made to reveal the effects of a particular color combination or cartographic technique.

In producing a virtual map we need not be concerned with the specifics of the technology with which the real map will be produced; it need not even have been invented yet. A similar idea is the distinction between hard-copy and soft-copy. First, a hard-copy map means one that we have produced that is actually going to last, using a stable medium in a form designed for viewing and use. Soft-copy maps came about with computer cartography. A soft-copy map corresponds to the work sheet in manual cartography. It is a map we can look at quickly to see if we like it, if we want to produce a hard-copy at all. With many geographic queries, a soft-copy map can answer our cartographic questions completely, meaning that we never need to produce a hard-copy map.

Soft-copy and virtual maps have given computer cartographers the option of using a design loop, a revolutionary tool for improving maps. The design loop is a system by which a map design is improved by interactive trial and error rather than by textbook learning. In manual cartography, map data are compiled, scales and projections are chosen, and a set of map types and symbolization techniques are selected. Finally, the cartographer invests the necessary hours in actually producing the map. With so much time invested, one is less likely to see flaws in the map design. If somebody then said "The title is too large and should be moved over 10 millimeters," there is nothing to be done, except to take note for the next time. If the map were a soft-copy, the cartographer could probably make the change in a matter of seconds and then ask the critic for any additional feedback. The ability to change design interactively in this way is the design loop. The design loop also frees the cartographer to learn from deliberate mistakes and allows experimentation with nontraditional designs.

Some soft-copy is indeed soft in that it is actually a temporary image, like a fax. Thermal printers have been used to produce map images, but are usually used only as a cheap intermediate stage before another form of hard-copy is generated. Most paper maps are in reality only longer-lasting versions of thermal prints. Paper tears, expands and shrinks, creases, fades, stains, and is combustible. Virtual maps, because they are technology independent, are virtually indestructible. Unfortunately, they are not yet practical enough to take hiking in the rain.

2.4 WORKSTATIONS FOR CARTOGRAPHY

The most important hardware revolution for cartography in recent years has been the incorporation of microprocessors directly into the display device. In the early days of computer cartography, only large computers did computer cartography, and users of the large computers were remotely connected via dial-in lines and stand-alone "dumb" graphics terminals. Special-purpose hardware devices such as digitizers and plotters were remotely connected at different locations and were often not available for hands-on use. Microprocessor technology has radically changed this state of affairs. Microprocessors took most of the capability coming from the large computer and put it into the terminal itself.

Over time, more and more functions and capabilities have been moved from the mainframe computer to the graphics terminal. The next logical step was to disconnect the graphics terminal from the mainframe, and move to a self-contained graphics device called a *workstation.* A graphics workstation can support multiple input and output capabilities and may have its own RAM, own disk, and usually even its own operating system. To gain access to mass storage, and to communicate with other workstations, the workstation is usually networked.

A typical workstation for cartography (Figure 2.7) may include a digitizing tablet, a mouse and a keyboard for input, a color and text graphic screen for output, and could be networked to plotters, laser-jet and ink-jet printers, and color-graphics cameras. The workstation concept is at the heart of the international graphics programming standard, GKS. In a workstation, any number of input and output devices can be supported as peripherals to the workstation. Graphics input devices can be locators, in which case they can point to locations on a map, or valuators, in which case they can simply indicate a value such as a dial or a joystick (Figure 2.8).

For menus, input can be choice, in which case the user selects from a set of options, or pick, in which case the user can select by pointing to a segment of the drawing on the screen, such as a box or an object. Graphics output devices are literally almost anything

Figure 2.7 A typical workstation. (Photo courtesy of Intergraph Corporation)

capable of producing a map. Using GKS, the concept of real and virtual maps finds expression in the terminology of the workstation, because a real map is simply a virtual map that has gone through the correct workstation transformations.

No discussion of hardware can be complete without a glance at the future and the new display technologies it holds. The one projection that will survive all the new technologies, however, is the workstation concept, which is versatile enough to incorporate any new hardware technology. Even portable workstations are now available, and breakthroughs in microprocessor technology have made the high-end microcomputer a workstation in all but name.

Some devices do, however, appear to have a cartographic future. First, a color screen shutter which can be coupled with polarized glasses to generate stereo screen images is available. Recent versions do not even require the glasses, but generate stereo nevertheless. In photogrammetry, remote sensing, and terrain analysis, this seems to be a device with some interesting prospects, as yet fully exploited. Second, the field of virtual imaging has also given us the helmet as an output device, in which the screen inside the helmet uses the entire field of vision of the viewer, and a glove or even a bodysuit as an input device. Maps that can be "virtually" explored (rather than virtual maps) offer an exciting future for cartographers.One vendor has introduced a head-worn display smalll enough to be worn while driving, flying, or performing other tasks. The ability to generate maps in real time so that they can be seen in peripheral vision is an exciting development and

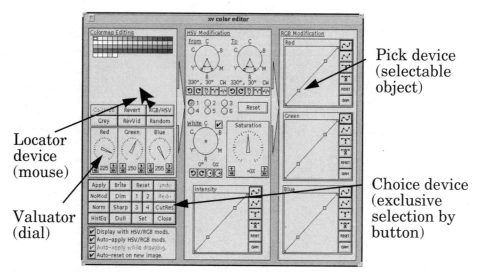

Figure 2.8 Different workstation Input devices supported by GKS.
(Example using xv by John Bradley. Used with permission.)

may rival the success of dashboard displays for vehicle navigation systems. Third, the linking of display technology with the real time capabilities of satellite-based navigation systems has great potential for cartography. While none of these technologies will influence how maps are made, the potential for influencing map use is extraordinary.

2.5 REFERENCES

Moellering, H. (1980). "Strategies for Real Time Cartography." *Cartographic Journal*, vol. 17, no. 1, pp. 12–15.

3

Software

3.1 SOFTWARE TRENDS

In Chapter 2, we examined the physical tools available for computer cartography. These tools, however, are merely expensive doorstops without the sets of computer instructions designed especially to allow the use of hardware for making maps. The computer instructions that interact with the cartographer and allow the manipulation of data, hardware, and design for computer mapping are collectively known as software. Computer cartographic software includes all the data, programs, packages, and interactive systems necessary for mapping.

Several general themes pertain to software. First, as computers have moved from large mainframes alone to a diversity of mainframes, minicomputers, workstations, and microcomputers, computer mapping software has tended to become available over the full range, so that capabilities which were once available only on mainframe computers are now available to all. Computer mapping has moved away from mainframe computers toward smaller and more efficient microcomputers and workstations with dedicated graphics capabilities. Software has reflected this trend, with a movement away from large special purpose mapping programs to small, general purpose programs.

The early computer mapping software programs other than the large commercial systems tended to be single-purpose, stand-alone programs. As software has moved away from the large computer companies and academia, the programs have become more integrated, sharing a user interface, common data structures, and capabilities.

Many automated mapping systems now are really just parts of much larger systems such as statistical analysis or surveying systems. In addition, computer cartography now overlaps with the field of GIS. GISs are integrated systems for the management and analysis of spatial data with computer mapping capabilities. Many systems designed for data presentation, for image processing, and for computer-assisted design, have some computer mapping capabilities.

In 1974, the International Geographical Union's (IGU) Commission on Geographical Data Sensing and Processing started to inventory the computer software for computer mapping, analytical cartography, and GIS (Brassel, 1977). Their three-volume report, published in 1980, was the first comprehensive survey of computer mapping software.

Volume 1 of the IGU report was a catalog of complete geographic information systems. Volume 2, entitled *Data Manipulation Programs,* surveyed general geographic software. Volume 3, which surveyed cartography and graphics (Marble, 1980), was the result of a survey that was conducted during 1978–1979, and as such gives a view of the types of computer mapping software available during the early years of computer mapping.

Many years have now passed since the inventory was started, and current inventories would be veritable dictionaries of systems. A few sources now survey, review, and evaluate software, and these are considered later in this chapter. Of current interest, however, is how the software of the IGU report was a portent of the contemporary distinctions between computer mapping software packages. The report will be examined by section, with a discussion of the current software in each category.

3.1.1 Geographic Information Systems

The report listed 38 complete mapping systems, defined as groups of programs with multiple mapping functions. In the list are several early prototypes of what later became GISs, which were initially developed as computer mapping systems. These prototypes show that until computer mapping software became fully integrated with database management systems, there was very little difference between computer mapping systems and GISs. The distinction is still vague, and indeed many GIS packages contain extensive control over the design and production of maps.

Software that falls under this category today, that is, systems with both data management capabilities and user controlled map display, include Arc/View and Arc/Info by ESRI, GIS Plus by Caliper Corporation, ERDAS Imagine (Map Composer) by ERDAS Incorporated, Intergraph's Modular GIS Environment, ATLAS*GIS by Strategic Mapping Incorporated, IDRISI by Clark University, and GRASS by the U.S. Army Corps of Engineers, among many others.

Typically, cartography within these systems is performed by building a *macro*. A macro is a master control language program that selects data; establishes map geometry, extent, and scale; specifies how symbolization is to be done (for example, color sets, line properties such as solid or dashed); places the symbols; and finally writes the map into a format compatible with an output device. The device driver is bundled with the software, so that a given set of output formats are provided. This is similar to the batch control of mainframe computers. Interaction is by directing the macro output to the screen, noting problems, and editing the macro file to correct it.

3.1.2 Map Design and Computer-Assisted Drafting Systems

Twenty-five listings were systems for data collection, editing, and development, while seven of the systems were designed to do automated lettering. Editing systems were essential, because in the early days manual digitizing produced map data that were very error prone, and all translations, rotations, and so forth, had to be conducted by stand-alone secondary programs. Twenty-seven systems performed basic drafting operations and design. These were systems that allowed you to draw in some interactive way on the computer screen.

Many of these systems evolved into the basis for an entirely new industry by themselves, the computer-assisted drafting and design (CADD) industry, which overlaps into architecture, facilities mapping, and engineering. More recently, these systems have spun off microcomputer and workstation versions that put basic drafting capabilities, such as the manipulation of high-quality linework and text, into the hands of the cartographer. CADD packages used extensively in cartography for data capture and especially editing are AutoCAD, VersaCAD, TurboCAD, and ArcCAD.

Inexpensive design industry versions of the original CADD systems have also evolved into specialized computer packages for the graphic arts and publishing industries. Only those with extensive graphics control, and import/export capabilities are of use in cartography. These include CorelDraw! by Corel Systems Corporation, FrameMaker by Frame Incorporated, Adobe's Illustrator, FreeHand by Freelance Software, MapGrapfix by ComGrafix Incorporated, Designer by MapGrafx Incorporated, Island Write/Paint/Draw by Island Graphics (Figures 3.1 and 3.2), and MacDraw II by Claris Corporation.

A common property of these systems is that they manipulate graphics as objects, that is, as a set of graphic primitives such as lines, text, and filled areas that can be *grouped* into a complex feature, termed a *segment* in GKS. As such, these graphics editors can be termed *object-based*. Active objects are usually highlighted by a box with corners drawn around the bounding rectangle for the object (Figure 3.1). Objects can be hierarchically grouped.

The package then allows a whole group to be rotated, translated, dragged, scaled, deleted, flipped and so forth, as a whole, rather than as constituent parts. Most systems also allow features to be layered, that is, some features can appear in front of or behind others. CADD systems allow snapping and copying of features between layers. Some packages allow erasure to "reveal" another layer, or translucence of layers. Others, for example CorelDraw!, feature raster-to-vector and vector-to-raster conversion and feature tracing.

Directly related to these *object-based* draw packages (not *object-oriented,* a different concept entirely) are the raster-only packages, commonly called *paint programs*. Paint programs allow the manipulation of images or bitmaps one pixel at a time only. Fills and text are immediately translated into bit maps, and object manipulation is not possible. But images can be resampled, sections can be cut out, areas can be color-increment shaded and so forth. Examples are Publisher's Paintbrush and PC Paintbrush V Plus by Zsoft (partially incorporated into the Windows Paintbrush), Superpaint by Claris Corporation, Dr. Halo by Media Cybernetics, and Adobe's PhotoShop.

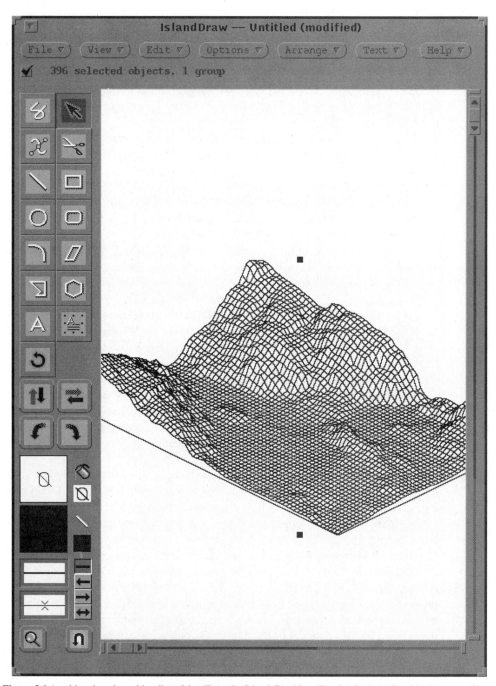

Figure 3.1 An object-based graphic editor: IslandDraw by Island Graphics. (Hewlett Packard Graphics language file imported from Surfer by Golden Software.)

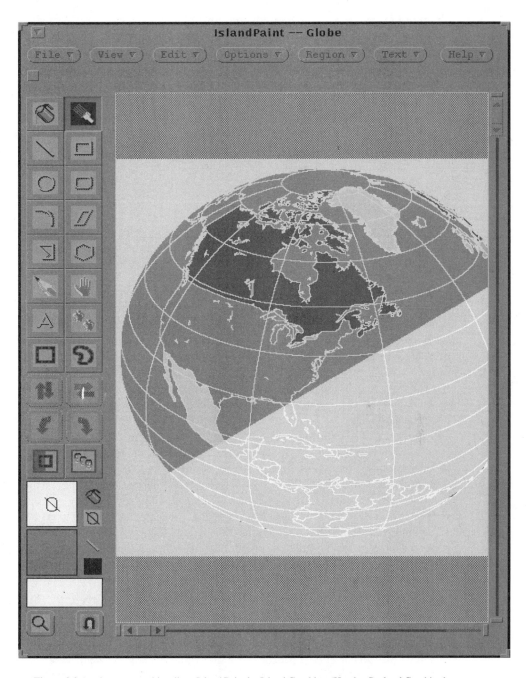

Figure 3.2 A paint type graphic editor: Island Paint by Island Graphics. (Hewlett Packard Graphics language file imported from MicroCAM.)

A key element that has arisen recently is the ability for these programs to share data. Plot files in many of the industry standard formats, such as PostScript, DXF (AutoCAD's Digital eXchange Format) or HPGL (Hewlett Packard Graphics Language), or in the International Standards Organization standard CGM (Computer Graphics Metafile) format can be moved between packages, even across operating systems and computer platforms. Because many stand-alone GIS and computer mapping packages can generate these plot files, this allows the major software packages to be used to generate maps, and then a CADD or design package can be used to add text, neat lines, special symbols and so forth. This is a particularly powerful combination, because the batch macro style of many of the basic mapping and GIS packages makes layout and design particularly tedious.

3.1.3 Systems for Map Displays of Data and Statistics

Nineteen of the packages displayed diagrams rather than maps, for example pie charts and histograms. Ten of the systems produced cartograms, which are maps with geographic space distorted to show an attribute other than area. Although not specifically cartographic, these packages have become another new industry, called presentation graphics. The vendors in this area have made the presentation of statistical and other information into a communications art and have produced some very effective products that take advantage of hardware devices such as projectors and slidemakers.

Products in this area break down further into statistical packages, presentation packages, digital Atlases, and animation packages. Statistical packages are those that are normally used to tabulate data and compute basic or advanced descriptive and parametric statistics. Some of these packages include additional commands which allow georeferenced data to be displayed, usually only as choropleth or point symbol maps. Among these packages are SAS Graph by SAS Institute and Harvard GeoGraphics by Harvard Graphics.

Presentation packages allow maps or graphics from other sources to be included within elaborate and highly attractive graphics backgrounds, so that presentation slides or posters can be made. Typically text and graphics can be laid out against a background that is supplied as a design template. Examples are FreeLance Graphics by Lotus, PowerPoint by Microsoft, Show Partner by IQ Technologies, and Charisma by MicroGrafx.

Digital Atlases are sets of ready-made maps grouped around a particular area. These packages have become extremely popular, and in many cases have replaced the basic printed Atlas. Because any map in the Atlas can be selected, cropped, and printed out, custom maps can be made, for example of highways in a given city. Special versions also exist for highway route planning. Examples are PCGlobe and PCUSA, Global Explorer and Street Atlas USA from DeLorme, and AutoMAP from R. Donnelley.

Animation packages can either take sets of maps as single frame images and play them in sequence to make animations or allow sophisticated editing and even "directing" of animation sequences. The packages often integrate with MultiMedia and Hypermedia authoring systems and are covered in detail in Peterson (1995). Examples are MicroMind Director and MediaDeveloper by Lenel Systems.

3.1.4 Feature Mapping Systems

Thirty-one of the systems in the IGU report did point and line mapping, producing maps of what is now called "cartographic spaghetti." These systems were often dedicated to specific databases, such as the World Data Bank, or a single type of plotter device. In 1980 many of the systems produced thematic maps, in particular the choropleth maps much in demand for statistical cartography. Here we see the split in technology, since 26 of the systems did area shading using grid or raster technology and 22 of them used vector technology. Almost all produced line printer output, printing and overprinting symbols to make patterns and shades.

Today, these packages represent the bulk of basic computer mapping programs. Key to these systems is the variety of data formats they support, the device drivers, and their particular ease of use for editing and design. Examples are MapInfo by MapInfo Corporation, Arc/View by ESRI, Atlas*PRO by Strategic Mapping, MacChoro II by Image Mapping Systems, GIMMS by Gimms Corporation, MapViewer by Golden Software, and FMS by Facility Mapping Systems.

These packages allow the capture of geographic data from basic sources (such as TIGER and DLG files), and can be used for annotation; symbol placement; line, color, and pattern control, and so forth. Many perform point symbol and choropleth mapping only, but some have more sophisticated choices, such as perspective views, stepped statistical surface maps, and cartograms. As mentioned above, a critical element is the ability to generate plot files that can be brought into other packages for layout and design.

3.1.5 Surface Mapping

Thirty-four of the systems did contouring and surface interpolation. Surface interpolation consisted of taking data collected as point samples, with an attribute such as elevation or the depth of a drill hole, and gridding the data for the purpose of making a contour map. Many of these systems have come about due to the needs of the oil exploration industry. Early versions of these systems produced very crude maps on a line printer using printed symbols, and lines had to be drawn in by hand afterwards. Few of them labeled the contours; instead, they printed legends along with the maps. The more sophisticated ones, packages like SURFACE II (Sampson, 1978), produced vector output, allowing smoothing and annotation, and had adequate user and reference documentation and support. Thirty-one systems were for three-dimensional mapping although most simply generated gridded perspective diagrams.

Three-dimensional surface symbolization, including contouring, is particularly well served by the software industry. Many of the major GIS packages, such as ERDAS and Arc/Info, have these capabilities. Other packages are MicroDEM by the U.S. Naval Academy, Surfer by Golden Software, Gridzo by Rockware, Inc., and PacSoft by PacSoft. High-end software for three dimensional surface representation overlaps with scientific visualization software and includes Dynamic Graphics, Precision Visuals, Silicon Graphics, and IBM's Data Explorer software.

The user interfaces for these packages vary considerably, and in the workstation software cases include entire fourth generation languages and program libraries for graphics, using interfaces to X-windows and in some cases proprietary graphic windowing systems. Many of the packages have advanced capabilities for using different interpolators, surface representation methods, contouring methods, and support grid, TIN, and other data structures.

3.1.6 Coordinate Conversion and Map Projections

Sixty-seven systems did either map projection transformations or other cartographic transforms such as translation, rotation, scaling, and distance measurement. These capabilities have found their way into most computer mapping software and rarely exist now as stand-alone programs. Operations supported by these routines include changing map projection, converting between coordinate systems, rubber-sheeting, affine transformations and many others.

Surviving computer programs under this heading exist for specific purposes such as coordinate conversion and education. Examples are the Geographic Calculator by Blue Marble Geographics, a special-purpose set of conversion routines to translate locations between datums, coordinate systems, and file formats; GeoLink by GeoResearch Incorporated, which reads and differentially corrects GPS data and writes the data directly into GIS and other mapping formats; and the General Cartographic Transformations Package, a set of FORTRAN programs available on the Internet and written by NOAA that translates between a list of map projections and ellipsoids.

In the education arena, the MicroCAM package, written by Scott Loomer of West Point and available through the Microcomputer Specialty Group of the Association of American Geographers, is unexcelled. The package runs on IBM-PC compatible microcomputers. It supports many map projections, point-symbol mapping, text placement, and so forth, and writes files into a number of standard formats.

3.1.7 Cartographic Data

Seventeen of the "systems" included in the original IGU report were actually data; in other words, they were not software as such but files containing data. As a function, the supply of data has sometimes remained attached to software; for example, several computer mapping software vendors offer ready-to-use data with their systems. Increasingly, however, data suppliers have emerged that supply redistributed government and other special purpose data sets. Among the leaders are ETAK, Geographic Data Technology, American Digital Cartography, and MapInfo.

Even where government agencies have produced comprehensive free cartographic data, private data suppliers have made an extensive business out of supplying digital maps, and recent years have seen data being produced by government and private partnerships. As a result, large sections of the surface of the earth are now available at many map scales, and both new and updated data become available all the time.

3.2 MAP SOFTWARE CHARACTERISTICS

3.2.1 Machine and Device Independence

One of the features listed in the IGU software survey was device dependency; that is, if a particular program required a specific type or brand of computer. In 1980, the level of hardware dependence was substantial, with much of the software not only requiring particular word sizes, computers, programming languages, and compilers, but also specific input and output devices made by only one manufacturer.

Although in some cases this level of machine dependence remains today, far more items of computer mapping software run on a variety of microcomputers, workstations, and mainframes. Many run under multiple operating systems, a particularly important feature. Software packages also support long lists of different input and output devices and even have installation menus and prompts that allow the support of multiple plotters, printers, digitizers, and other peripherals. Hardware vendors have realized the advantages of this, and they often manufacture devices so that they emulate the protocols of the more popular devices.

Unfortunately, the trend toward independence of machine and software is not yet complete. This means that as the market changes, devices already purchased are sometimes left without applications software. When purchasing software, it is more important to match the software against mapping needs rather than against your existing hardware. Similarly, if hardware is available, software that supports the most general and highly supported devices and standards is preferable.

3.2.2 Transportability

Machine and device independence and the ability to move software and data between computers are functions of the transportability of software. Transportable software can move between computers, operating systems, and devices with no more work than recompilation. Some computer programming languages are not transportable, with major variations from compiler to compiler and version to version. Software written in these languages will eventually disappear as programming environments change. Even operating system upgrades can make applications unusable, let alone the major changes that come about when new computers are installed.

The principal way in which transportability of cartographic software is ensured is by separating the graphics functions from the language. Even graphics functions change over time, and the only way to ensure against the loss of graphics capabilities is to use graphics standards, which will be supported over time.

Within computer cartography, many vendors have been slow to write standards-based software, meaning that many computer mapping programs remain tied to a particular set of devices, operating system, or working environment. As highly portable languages such as C and Pascal form the basis of more mapping systems, and as graphics standards such as GKS, CORE, and PHIGS are used to build new software, users will be freer to transport software and even maps between computers and through operating systems.

In the last few years, a convergence of operating systems upon a set of loose standards widely known as Open Systems has significantly increased the transportability of software. Programs that once ran only on one platform or under one operating system, such as the Macintosh, now cross over platforms. At the same time, operating systems have become less attached to platforms. UNIX and Windows NT are likely to support a large number of platforms.

Unifying elements have been the OS/2 operating system, the Motif Graphical User Interface, and the X-Windows system. Some systems even allow the coexistence of more than one operating system on one microcomputer or workstation, significantly expanding the pool of software that suits a particular computer. Without any doubt, the ties between software and a single working environment have quickly eroded to the extent that they rarely constitute a problem as far as cartography is concerned any more.

3.3 TYPES OF SOFTWARE

3.3.1 Software for Workstations

Computer mapping software for large computers is mostly that which runs on minicomputers and mainframe systems. Among the many vendors are Synercom, Intergraph, and ESRI. Large computer systems often require special hardware that can be purchased only though the software vendor. Many vendors manufacture their own workstations rather than depending on other manufacturers, and they supply software and hardware bundled together in "turnkey" or ready-made packages. Most minicomputers and mainframes have been used for graphics and mapping, but the more popular types are DEC's VAX and microVAX computers, and minicomputers by Prime, IBM, and others. These use a large variety of operating systems, such as DEC's VMS and Prime's PRIMOS.

The lower end of the minicomputer market, however, has been completely overtaken by the workstation market. Leading workstation manufacturers are Silicon Graphics, Sun, Hewlett-Packard, and Digital Equipment. Although the workstations use a variety of systems, most use the UNIX operating system. Languages common to many of these systems are Pascal, FORTRAN 77 (in which a large amount of computer mapping software is written), and, increasingly, C. These languages and their compilers are becoming standardized over the various hardware environments due to the efforts of the standards organizations, especially the American National Standards Institute (ANSI). The use of an ANSI standard version of a language means that a program written on one computer and sent to another by tape, file transfer, or network link will compile without problems arising from the specifics of the machine on which the compiler is running.

Most of the major software packages for computer mapping and GIS have been ported to the workstation environment. In many cases, the first ports were simply a translation of existing software. Later versions, however, have been completely rewritten, usually in C and using X-windows, to support the full set of functions available in the workstation environment. Erdas Imagine, AutoCAD, Arc/Info, Arc/View, and CorelDraw! are examples of software that have recently undergone this important translation.

3.3.2 Software for Microcomputers

There are a great many computer mapping systems that run on microcomputers. It should be recognized, however, that microcomputers vary just as much as workstations and minicomputers in speed, size, capability, and peripherals. The first microcomputer mapping software appeared after the popularization of the microcomputer by Apple, and then by IBM during the very early 1980s. In the years since the origin of microcomputer software, the microcomputer has become infinitely more powerful, to the point where the boundary between PCs and RISC workstations becomes very blurry. Many vendors have produced software to take advantage of this power. Like the industry itself, which has produced "shakedowns," some vendors have produced fine products and then disappeared completely.

The microcomputer programs surveyed above are simply the tip of the iceberg. Information about new and existing software can be found as software reviews in *The Professional Geographer,* published by the Association of American Geographers; and in *Cartography and Geographic Information Systems* (formerly the *American Cartographer),* the journal of the American Cartographic Association of the American Congress on Surveying and Mapping. Computer journals, such as *Byte, PC World,* and *InfoWorld,* and periodicals and trade journals, such as *GeoInfoSystems, GPS World, Government Technology, Computer Graphics World,* and *GIS World,* all carry product reviews and announcements, advertisements, and other product and supplier information.

Some sources can be considered "lists of lists." Among these are *Resource Guide for GIS* (Supplement to Government Technology, 9719 Lincoln Village Drive, Suite 500, Sacramento, CA), *International GIS Sourcebook* (GIS World Inc., 155 E. Broadwalk Drive, Suite 250, Fort Collins, CO 80525), and *Environmental GIS Applications Guide, 1993* (Digital Equipment Corporation, 444 Whitney St., Northboro, MA 01532).

Three books reprint software reviews and list software:

Cassettari, S. (1993). *Introduction to Integrated Geo-Information Management.* London: Chapman and Hall. Lists about 150 programs, classified by broad category (for example, GIS, CAD) and platform (PC or Macintosh).

Clarke, K. C. (1990). *Analytical and Computer Cartography*, 1st ed. Englewood Cliffs, NJ: Prentice Hall. Appendix A lists 68 software products with vendor names, addresses, and phone numbers, plus references to reviews of the software in other journals. The appendix was not included in this edition because it is now too long a list.

Dent, B. D. (1993). *Cartography: Thematic Map Design* (3rd ed.) Dubuque, IA: W. C. Brown. Appendix reprints reviews from journals of several programs.

3.3.4 Software for Software

Software development tools are pieces of software designed for writing computer mapping systems. They are designed to assist programmers, or in some cases advanced users, in producing their own software for automated mapping. Early software development tools were simply large collections of FORTRAN subroutines to support mapping and graphics. Among the suppliers of workstation and microcomputer graphics toolboxes are Precision Visuals (DI-3000 and GKS-2000), GSS*GKS by Ematek, UNIRAS, and Halo. Many of these packages are devoted to a particular device and a particular graphics card, such as the EGA graphics on an IBM-PC. Graphics support is usually provided with program-callable functions which plot points, lines, and areas. Other functions allow color selection; plot symbols and text; and control other graphics capabilities. Although some packages support their own macro language, most assume at least a working knowledge of a computer programming language such as C or Pascal.

Some microcomputer versions of programming languages have a limited support for graphics, by using simple extensions to the language using plotting functions. Very often these also are highly restricted to the particular graphics card in use, the size of the screen, the color mode in use and so forth.

In the chapters to come, a particular software tool, the GKS graphics standard, and the various bindings of this standard to the C language are discussed. The strength of this tool is its device independent nature, that is, the fact that the production of maps on a specific device is controlled by a separate piece of software called a device driver and not by the hardware used. The device driver interacts with more generic software, the standard, which does most of the map production. This system is highly flexible and is well suited to the demands of both computer and analytical cartography.

3.4 SOFTWARE RESOURCES

Computer cartography is a rapidly changing field. Anyone who purchases software can almost be certain that the next year will bring more and better alternatives. Computer mapping software can also be expensive, although prices are continuing to fall. An alternative source of software can be found in the user group and in shareware. Shareware is software supplied at cost or at a nominal fee to anyone interested. In the United States a major clearinghouse for shareware is PC-SIG (PC-SIG, 1987). Software also circulates free on many computer bulletin boards and is published in journals such as *Byte* and *The C Users Journal.* An immense quantity of software is available via the Internet, covered in Chapter 6, and it can be located and retrieved using the Internet tools discussed there.

Within the field of geography, the Association of American Geographers' Microcomputer Specialty Group also distributes shareware. This group's newsletter also reviews software packages. Many databases and computer programs are available from these sources, although their quality, scope, and languages vary considerably. These sources are excellent tools for getting a close look at the capabilities of computer mapping systems before spending money on a more comprehensive mapping system. Cartographers in particular, especially those capable of producing computer cartographic software, should make an effort to support these organizations.

3.5 REFERENCES

Brassel, K. E. (1977). "A Survey of Cartographic Display Software," *International Yearbook of Cartography*, vol. 17, pp. 60–76.

Marble, D. F. ed., (1980). *Computer Software for Spatial Data Handling, Vol. 3: Cartography and Graphics*. Ottawa: International Geographical Union, Commission on Geographical Data Sensing and Processing.

PC-SIG. (1987). *The PC-SIG Library*. Sunnyvale, CA: PC-SIG Inc.

Peterson, M. (1995). *Interactive and Animated Cartography*. Prentice Hall Series in Geographic Information Science, Englewood Cliffs, NJ: Prentice Hall.

Sampson, D. (1978). *The Surface II Graphics System*. Lawrence, KS: Kansas Geological Survey.

PART II
The Representation
of Cartographic Data

Cartographic data form the base structure around which analytical and computer cartography are constructed. Five chapters in this part cover the issues related to capturing cartographic data and structuring the information for representation in digital form. Chapter 4 covers geocoding, the conversion of cartographic data from the real world and from paper maps into digital objects. Chapter 5 covers the major means of structuring the information physically onto storage media such as disks and tapes. The most popular distribution formats are covered in detail. The section is conceptually linked to the specifications of the Federal Information Processing Standard for spatial data. Chapter 6 covers how to access spatial data directly and indirectly using the Internet and other means. Chapter 7 deals with how to store the data in simple C language data structures within computer programs. Chapter 8 reviews the various conceptual data models that have been advanced for use with spatial data, and again gives C language examples for some of them.

4

Geocoding

4.1 GEOCODING AND COMPUTER CARTOGRAPHY

Geocoding is the conversion of spatial information into computer-readable form. As such, geocoding, both the process and the concepts involved, determines the type, scale, accuracy, and precision of digital maps. An important aspect of geocoding is that effective geocoding requires an understanding of some basic geographic properties underlying geographic data. Unfortunately, it is quite easy to convert information into computer-readable form as data, but not all data are information, and all too frequently analytical cartography drowns in a sea of meaningless and unusable numbers. Most essential for analytical and computer cartography is that the information should be about geographic phenomena. The role of the cartographer is to subdivide the broad landscape elements of geographic interest into smaller units more suitable for mapping, termed here *cartographic entities*. The cartographic entities can then be photographed, measured, sampled, or surveyed and then entered into computer mapping systems via *geocoding* methods to become *cartographic objects*.

Geocoding is the first stage of computer cartography and is done for two reasons. First, we can geocode with a specific mapping purpose in mind, for example, capturing lines that are going to be parts of polygons to be used to produce a choropleth map of Western Europe. Second, we can do general purpose geocoding; in other words, we can collect every piece of information we can about a specific area and assemble it for service in many possible cartographic contexts. The geocoding of general purpose cartographic data has great potential importance, just as accurate mapping of the country by topographic survey was important. The agencies performing this type of data collection and encoding have traditionally been government agencies, and we are fortunate that these agencies have often been at the forefront in making geocoding effective and efficient.

We can also geocode at two different levels. First, we can simply convert the graphic elements of a map into numbers so that we can reproduce the map using the methods of computer cartography. Many geographic databases consist of geocoded data in this form. Alternatively, we can encode important topological information about the data we are geocoding. Increasingly, as GISs become the principal users of geocoded data, the encoding of topology becomes essential to the use and survival of cartographic data sets.

To understand this distinction, a good example is the symbolization of roads and rivers. If the digital map consists of a road and a river that cross, a nontopological approach would be to plot the two lines one on top of the other. If the topology is encoded, we could recognize that either the road crosses the river on a bridge, in which case we draw a road symbol only and break the river symbol, perhaps adding the symbol for a bridge. Alternatively, the river could cross the road, implying an aqueduct or tunnel. Critical to this instance is that both lines, the river and the bridge, must be broken to include the feature at the intersection as part of their geometry. The importance of topological geocoding is considered later. First, however, we will return to the idea mentioned above, that effective geocoding comes from a clear understanding of the fundamental geographic properties and their manifestations.

4.2 CHARACTERISTICS OF GEOGRAPHIC DATA

If the purpose of geocoding is to encode the fundamental characteristics of geographic data digitally, then we must understand what those characteristics are before we start designing strategies for capturing them. The characteristics of geographic data are summarized in Figure 4.1.

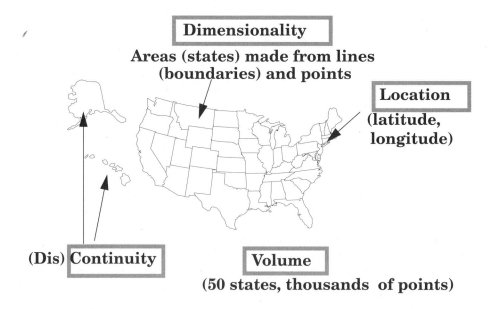

Figure 4.1 Some characteristics of geographic data.

4.2.1 Location

Fundamental to geographic data is the attribute of *location* on the earth's surface. Although we have to use a third dimension to describe elevation on the earth, the two dimensions of location on the plane or sphere are probably *the* basic geographical property. Normally, *x* and *y* values represent latitudes and longitudes, but often in geocoding we assume a map projection, such as a transverse Mercator, and we use coordinate systems such as Universal Transverse Mercator or State Plane to give locations. Occasionally, polar coordinates are used, giving an angle sometimes clockwise from north, and a distance.

Most coordinate systems for use with computer mapping are based on cartesian coordinates. This implies that the axes of the two directions, such as eastings and northings, are orthogonal, that is, they are at right angles to each other. This allows us to specify a location in space by referring to a pair of coordinates (x, y), or an easting and a northing. Leaving out one or both of these means that a location is simply undefined. Location is therefore the most fundamental characteristic of both cartographic and geographic data.

4.2.2 Data Volume

A second fundamental characteristic of geographic data is *high volume*. Computer science traditionally has dealt with small databases by cartographic standards. Cartographic and geographic databases contain thousands or sometimes millions of data elements, an amount that is directly related to the scale of the data and the extent of the map area.

A typical application in remote sensing uses seven one-byte bands of attributes and image arrays typically 512 by 512 in size. This represents a quarter of a million points for which we have multiple pieces of information all required for a single display. Many cartographic data processing problems are generic problems of large data sets. As a result, in computer cartography we often have to deal with memory constraints and the efficiency of our data structure.

Fortunately, the cost of storing data has decreased dramatically. Even on small computers, new storage methods have increased available memory from kilobytes to gigabytes within only a decade. The effect has been to change the emphasis from simple storage volume to storage access time as the primary volume-related consideration, although memory constraints will remain a factor.

4.2.3 Dimension

A third fundamental characteristic of geographic data is *dimensionality*. Traditionally, cartography has divided data into *points, lines,* and *areas.* Somewhat related is the concept of level of measurement. Levels of measurement are divided into *nominal, ordinal, interval,* and *ratio.* These two divisions have formed the basis of at least two classifications of mapping methods and cartographic data and are considered in more detail in Chapter 8.

4.2.4 Continuity

A fourth major characteristic of geographic data is *continuity*. Some map types, such as contour maps, assume a continuous distribution, while others, such as choropleth maps, assume a discontinuous distribution. Continuity is an important geographical property. The best example of a continuous variable is probably surface elevation. As we walk around on the earth's upper surface, we always have an elevation. There is no point where elevation is undetermined.

In real terrain, there are very few exceptions. Vertical overhangs and cliffs do indeed have areas where elevation is locally undetermined, but on the whole, elevation as a geographic distribution is continuous. Continuity does not always apply to statistical distributions. For example, tax rates are a discontinuous geographic variable. A resident of New York has to pay the state personal income tax, but by living just 1 meter inside Connecticut, a person will not pay that tax. In such an example, the tax rate is a discontinuous geographic variable because on the boundary line, the tax rate is undefined. It was once somewhat facetiously suggested that the only truly discontinuous geographic variables were tax rates and road surfaces, as anyone who has paid New York taxes or driven into the city of New York knows.

In addition, geographic continuity is an important property. Space classifications by areas must be exhaustive for continuity, that is, there should be no holes or unclassified areas. Similarly, for a set of categorical attributes reflecting a map, the set should contain all the objects found on the map, without any "other" or "miscellaneous" categories.

4.3 FUNDAMENTAL PROPERTIES OF GEOGRAPHIC OBJECTS

So far we have discussed the characteristics of geographic data. We should now consider some of the fundamental underlying properties that shape geographic and cartographic objects. Although characteristics of geographic data influence computer cartography, the implications of the fundamental properties are closer to the concerns of analytical cartography. A full understanding of the properties of geographic objects allows more effective geocoding, provides for the correct use of cartographic data structures, and facilitates the use of cartographic data transformations. The properties are illustrated in summary form in Figure 4.2.

4.3.1 Size

A basic underlying property is *size* and its characterization in measurement. Most geographic phenomena can be measured directly, for example by survey or air photo. A point has the measured aspects of location (x, y), adjacency, and elevation. A line has length, direction, connectivity, and "wigglyness." A polygon has topology (whether there are holes or outliers), area, shape, and boundary length, as well as location and orientation. A volume has topology, continuity, surface slope, surface aspect, surface trend, structure, location, and elevation. Most of these properties are comparatively simple to measure if the cartographic data are geocoded. Some are extremely difficult to measure in the real

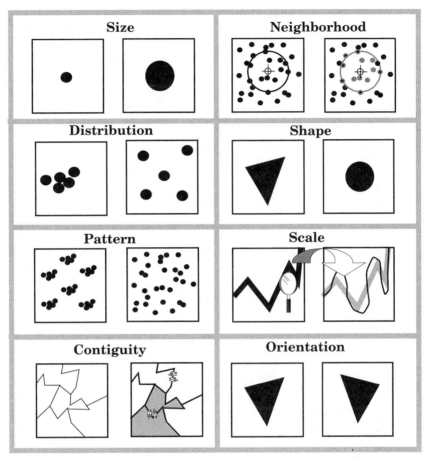

Figure 4.2 Fundamental properties of geographic objects.

world and as such can only be analyzed using the data abstractions of mapping and representation as cartographic objects. These measurements can usually be implemented by simple algorithms, some of which are given in the following chapters

4.3.2 Distribution

Another fundamental property is *distribution*. Density is a measure of the distribution of a phenomenon across space. Density can be computed by counting cartographic objects or their attributes over a set of geographic units, such as a grid or a set of regions. The density of a geographic phenomenon has a great many implications not only for how we measure and geocode it, but also how we can generalize it. This influences our decision on map coverage, and how we symbolize it on a map.

4.3.3 Pattern and Orientation

Another fundamental property of geographic objects is *pattern.* Pattern is actually a characteristic of distributions and is a description of their structure. Pattern can be thought of as a lack of randomness. A "first law of geography" is that anything that is of geographic interest lies at the intersection of between two and four maps, photos, or images. This relates to distribution via the sampling theorem.

The "second law of geography," with apologies to Tobler, is that everything is related to everything else, but *near* things are more related than others. In other words, if two things are close to each other, they are more likely to be similar than if they are separated by a long distance. This implies that proximity and adjacency imply a stronger interrelation between geographic objects, and that the strength of the relationship can be measured.

The simplest way in which this "nearness" phenomenon can be seen and measured is by repetition. Those relationships that repeat themselves at distances of less than half the size of the map result in patterns, implying that repetition is very important for pattern. For simple point distributions we can measure randomness, or lack of randomness, by using the nearest-neighbor statistic.

A test for pattern goes beyond simple measurement, however, because to observe pattern we need a model or description of the pattern we wish to find. In remote sensing and image processing, we can pass "templates" over a continuous distribution looking for a match of the discontinuous feature we wish to detect. This approach works even if the image is obscured by error or atmospheric effects. Patterns repeated through scales are also of interest and can be called "self-similar."

Probably the simplest pattern is direct repetition. Another basic property of objects distributed in space, however, is their *orientation.* In point distributions, elongated distributions or asymmetrical patterns show orientation, lines have obvious directions, areas are sometimes rotated and scaled equivalents of each other, and surfaces have dip and aspect, all implying orientation.

4.3.4 Neighborhood

The *neighborhood property* is, along with pattern, one of the most definitive of geographic properties. If pattern is the repetition of an attribute over space, the neighborhood property defines how the property varies over space. A key aspect of the neighborhood property is that variation takes place with distance, so little separations mean similarity, and big separations mean dissimilarity. Geographic studies often examine the relationship between some geographic phenomenon and distance.

Usually we see the neighborhood property as a distance function, a mathematical expression of the relationship between a geographic property and distance. Within geography, this distance function has been characterized and measured using tools such as the autocorrelation function, spatial interaction models, distance decay models, and the variogram. These functions express mathematically exactly what the second law of geography states.

4.3.5 Contiguity

Contiguity is another important geographic property. Contiguity is the property of being related by juxtaposition, that is by a sharing of a common boundary. Contiguity, therefore, is one of the geographic expressions of topology. The best example of contiguity is the sharing of a common boundary. Political geographers may be interested in the length of the border of Poland, perhaps even the geometric shape of Poland, but the main item of geographic interest would be a list of nations with a common boundary. Similarly, in land cover terms, we may be interested in the land uses that are most likely to be found around lakes. Whether the answer is beach cottages or swamps, we have a distribution of direct geographic interest

Contiguity is expressed in many ways. We can define it in terms of shared boundaries, in which case we can measure the number and lengths of shared boundaries. Within networks, contiguity is referred to as connectivity. For example, a network may contain connected links between nodes. In the sense of the network, the nodes are contiguous; that is, spatially you can get there from here in one step. Also, often we think of contiguity in terms of pixels or a grid structure. In fact, we do this so frequently that we even have special terms for it: four-cell and eight-cell contiguity (Figure 4.3). The center pixel for a neighborhood is called the kernel, and the contiguous pixels are those that share a direct common boundary or a corner.

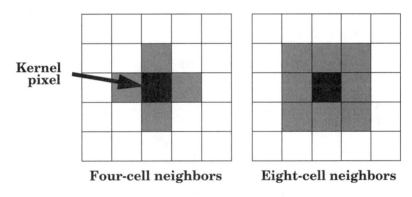

Kernel pixel

Four-cell neighbors   **Eight-cell neighbors**

Figure 4.3 Contiguity of pixels.

4.3.6 Shape

Shape is another very geographic property. Shape is a difficult property to measure directly, so most shape measures are really measures of the level of correspondence between shapes. Some, however, are graphic with fewer dimensions. An excellent review of shape measurement is that by Pavlidis (1978). Scientists have measured the shapes of many phenomena. For example, geologists have measured the shape of Pacific atolls in the ocean, and biologists have measured the shape of cells and the wings of butterflies. Many geographers are interested in the shape of cartographic objects, such as the shapes of congressional districts or geomorphological features.

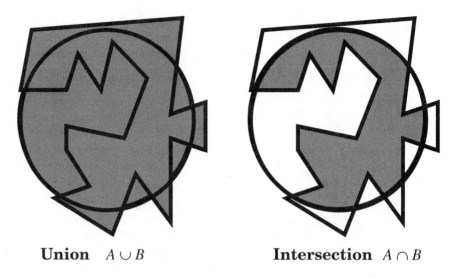

Union $A \cup B$ **Intersection** $A \cap B$

Figure 4.4 The Lee and Sallee shape measure.

It is best to illustrate the property of shape by example. A particularly simple shape measure is that of Lee and Sallee (1970), illustrated in Figure 4.4. This measure chooses a point somewhere inside the shape and draws a circle with the same area as the shape. The two figures are then overlain, forming three types of regions, an overlap or intersection, and then remaining parts of each of the two shapes. In set theory, the shape and the circle can be termed A and B. Then the Lee and Sallee shape measure is expressed as

$$ s = 1 - \frac{A \cap B}{A \cup B} $$

Notice that for a circle, $A \cap B$ and $A \cup B$ are both equal to 1. Subtracting $\frac{1}{1}$ from 1 gives zero. So the base measurement is zero, and the shape measure compares shapes with a circle. Note, however, that the shape number depends on the area of the chosen "comparison shape," the point at which the two figures are overlain, and the orientation. The same measure could be used to compute the resemblance between any set of shapes, such as the resemblance of the remaining 49 state boundaries to the shape of New York. There are many other shape measures. Shape, however, is a very complex property. For example, the measure described above could give the same number to an infinite number of different shapes. An inversion of the method, for example, to perform recognition, would be impossible.

4.3.7 Scale

The final fundamental geographic property is *scale.* As far as cartography is concerned, the property of scale is the most distinctive feature of things that are geographic. The simplest expression of scale is the representative fraction, the ratio of distances on a map to the same distances in that part of the world shown on the map. There is a limited range of scales over which cartography has an interest in a phenomenon. If instead of thinking about things at a particular scale we think about things through scales, then one of two things can happen. Either objects become more clear at certain scales, or they never change with scale. These two subproperties are called scale dependence and scale independence.

If things do not change with scale, they can be modeled using fractal geometry. Fractal geometry has dimensions that are not whole numbers. If we go back to our original geographic characteristic of dimension, we can consider the point, line, area, and volume as special geometric cases. In fractal geometry, lines have dimensions between 1.0 and 2.0, and areas and surfaces have volumes between 2.0 and 3.0. For example, a line on a map is represented by a geometrically infinitely thin cartographic object, yet it can only be symbolized by covering an area on the map with color. Scale dependence means that there is an appropriate scale for making a map. Analytical cartography can contribute by determining this scale objectively.

4.3.8 Measurement and Fundamental Properties

The purpose of geocoding from an analytical cartographic standpoint is to encode digitally the fundamental properties of geographic entities with the objects so that their later analysis is possible. Encoding these properties, usually as geographic relationships between objects, can sometimes be performed during the direct data capture process, but this encoding is most often captured by processing geocoded data using computer algorithms. Examples of quantitative measures used to represent each of the fundamental properties are summarized in Figure 4.5. Many of these measures and the algorithms for their computation require, in addition to explicitly locational data, the geocoding or the computation of the topological relationships between objects.

4.4 GOALS OF GEOCODING METHODS

Geocoding seeks many conflicting goals. Each should be kept in mind before geocoding cartographic data, and the computer cartographer should be realistic about which goals are at odds with each other.

4.4.1 Minimize Labor Input

Because one of the major sources of error in the geocoding process is human error, and because labor costs are usually high, especially for semiautomatic digitizing, an important goal for geocoding is to minimize the amount of manual labor involved. The manual component for converting existing paper maps is high and can multiply unexpectedly. Chrisman (1987) reported manual digitizing times of eight hours per 300 polygon soil

	POINT	LINE	AREA	VOLUME
SHAPE	Feature type	Curvature	Shape measure	Dimension Resemblance to figure (e.g. cone)
CONTIGUITY	Link	Intersection	Shared boundary	Shared face
ORIENTATION	Of cluster or pattern	Bearing Trend	Of axis Of pattern	Dip, drift, trend, aspect
SIZE	Number	Length	Area	Volume Surface area
SCALE	Range at which object is a point	Range at which object is a Line	Range at which object is an area	Range at which object is a volume
NEIGHBORHOOD	Set of nearby points	Connected lines. Lines within a range	Contiguous areas Area within a range Connected areas	Adjacent voxels Overlapping volumes Shared faces
PATTERN	Pattern matching Fourier analysis	Curve measures Fractal dimension	Shape distribution Description	Fourier power spectrum Trend surface
DISTRIBUTION	Standard distance Nearest neighbor number Autocorrelation	Line density Length, Intersection frequency	Coverage Autocorrelation	Variogram

Figure 4.5 Quantitative descriptors of the fundamental properties of geographic objects.

map sheet, with an additional four hours of editing, even using software designed explicitly for the reduction and detection of digitizing errors, with automatic topological correction and automatic end-node snapping. Chrisman pointed out that careful attention to the editing capabilities of digitizing software, coupled with consistency checking, can allow even a low-cost microcomputer workstation to produce high-quality, accurate, digital cartographic data.

For example, we may seek a digital land-cover map at 1:500,000. One approach would be to draw a grid with squares of 20 mm, representing 5 km on the final grid after scaling, over four 1:250,000 land-use and land-cover maps from the U.S. Geological Survey. We could then go grid cell by grid cell writing down the land-use category at each intersection. We could then type the land-use category numbers one by one into a computer file. Assuming that we can draw the grids and write down the numbers at the rate of 200 per hour, and can enter the data at the same rate, we would need about 400 hours, or 10 weeks of full-time labor, to geocode the map, without any checking.

Although in actuality geocoders become faster at a task with experience, anything we can do that makes data entry easier, we should do. In many cases, simple steps can be taken. Digitizing tablets that are adequately lit, and are repositionable, especially so that the user can sit down, can save substantial amounts of labor time, because much work is necessary in going back over errors and in cross checking. Immediate video and audible feedback for errors can save starting over after a string of errors, as also can direct editing, off-line control, and a quiet, nondistracting workplace.

More substantial steps can also be taken to reduce labor. Many steps are totally unnecessary, such as the writing down on paper in the example above. If in doubt, the best way to estimate the time for each step of the digitizing process is to conduct a small test on a pilot data set, taking it through the entire procedure. The time can then be multiplied up to get overall estimates.

4.4.2 Detect and Eliminate Errors

Many early geocoding systems had only limited editing capabilities. They allowed data entry, but error detection was by batch processing and correction was by deletion and re-entry. Anything we can do in the geocoding process that reduces errors, or that makes errors easily detectable, we should indeed do. As an absolute minimum, data for lines and areas can be processed automatically for topological consistency, and any unconnected lines or unclosed polygons can be detected and signaled to the user.

The easiest way to avoid errors in geocoding is to make sure that they are detected as soon as possible and then to make their correction easy. Video display during digitizing and audio feedback for error messages are essential. Software should spell out exactly what will happen in the case of an error. A common geocoding error is to overflow a hard or floppy disk while digitizing. Some software continues to accept data as if nothing is wrong, and gives a "disk full" message only when you exit from the digitizer software.

Some easy-to-detect errors are slivers, spikes, inversions, and disappearing nodes (Figure 4.6). Scaling and inversion errors are usually due to an incorrect digitizer set-up procedure, that is, they are systematic errors caused by incorrectly entering the control points for establishing the map geometry. Spikes are random hardware or software errors in which a zero or extremely large data value erroneously replaces the real value in one of the coordinates. Spikes are also sometimes known as zingers. Errors in topology, missing or duplicate lines, and unsnapped nodes are operator errors (Ward and Phillips, 1987).

Errors which are more difficult to detect take more effort. Some rules of thumb for the most difficult errors are as follows:

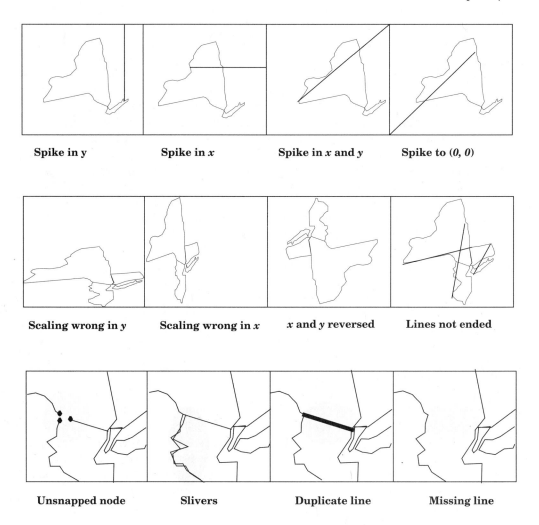

Figure 4.6 Some easy-to-detect geocoding errors.

- ■ If something looks wrong on the display, it is probably wrong in the data.
- ■ Believe the locational data files before the display.
- ■ Believe the source map before your version of it.
- ■ Stopping or having another person work on geocoding often is a source of errors.
- ■ Postponing a correction gives the correction only a 50% chance of ever being made unless postprocessing for errors is available.
- ■ Sloppy geocoding makes unreliable and erroneous maps.

In the Spatial Data Transfer Standards, a distinction is made between positional and attribute accuracy and logical consistency. Positional and attribute accuracy can be tested and measured by direct comparison with the source. Very useful aids in detecting errors are overlay plots at the same size and scale as the source map, especially if they can be superimposed on the original. There is no substitute, however, for systematically parsing through the geocoded data looking for discrepancies. Often, plotting the data becomes a useful aid because unplottable data often have bad geocodes. Similarly, attempting to fill polygons with color often detects gaps and slivers not visible in busy polygon networks. The best check for positional accuracy is a check against an independent source map of higher accuracy.

A data set that is correctly geocoded both positionally and with attributes is not necessarily logically consistent. Logical consistency can be checked most easily for topological data. Topologically, data can be checked to see that all chains intersect at nodes, that chains cycle correctly in a ring around polygons, and that inner rings are fully enclosed within their surrounding polygons. Otherwise, attributes can be checked to ensure that they fall within the correct range and that no feature has become too small to be represented accurately.

4.4.3 Optimize Storage Efficiency

Data contains two parts: volume and information. Geocoding often seeks to eliminate volume while retaining information. We need to know what parts of the data are redundant, what can be left out, and what must be kept.

The ideal storage method for geocoded data saves only the minimum, the distilled information. Redundant information, however, is often a necessary part of geocoding. A straight state boundary on one map projection may become curved on another and would need many points to make it look curved, many more than the "optimal" two necessary for a straight line. In addition, storage efficiency depends entirely on which data structure we will be using to map the digital data or how we will move the raw geocoded information between systems or applications.

4.4.4 Maximize Flexibility

A critical goal is to optimize flexibility. The most elegant data structure in the world is worthless if the author is the only one who can read it or if wants to read it because it is so obscure. We need data structures and storage methods that allow analyses we did not anticipate when we geocoded the data originally. These unanticipated data uses are not the exception, they are the norm. This flexibility is the very aspect of computer cartography that has allowed the development of geographic information systems.

As soon as data exists in digital form new uses for the data will be discovered, each of which will place a new set of demands on the chosen data structure. A measure of the flexibility of a data structure is how well it holds up to these unanticipated demands, rather than the demands of the original uses for which it was designed. Geocoding, therefore, should not impose upon the data restrictions on accuracy, precision, and reliability.

4.5 LOCATIONAL GEOCODES

Geocoding uses a map coordinate system. Coordinate systems can be standard or arbitrary. There are many reasons to avoid arbitrary referencing systems, among them incompatibility, lack of ability to document the alternative systems, inability to adapt to new levels of precision, or the assumption of a specific map projection. An important factor in choosing a coordinate system with which to geocode a map is its universality. A good coordinate system works worldwide; it is simple, accurate, precise, terse, and adaptable. Unfortunately, the problems of global coordinate systems are much like those of map projections; that is, not all goals can be served at once. As a result, we have several established systems for specifying locations on the earth's surface.

4.5.1 Geographic Coordinates

Many global data-bases record locations using latitude and longitude or geographic coordinates. Latitude and longitude are almost always geocoded in one of two ways. Latitudes go from 90 degrees south −90 to 90 degrees north (+ 90). Precision below a degree is geocoded as minutes and seconds, and decimals of seconds, in one of two formats: either plus or minus DD.MMSS.XX, where DD are degrees, MM are minutes, and SS.XX are decimal seconds; or alternatively, as DD.XXXX, or decimal degrees. Longitudes are the same, with the exception that the range is −180 to +180 degrees. In the second format, degrees are converted to radians and stored as floating point numbers with decimal places.

It is especially important that we record how each map has been geocoded if geographic coordinates are used. Maps look particularly strange when decimal degrees are taken for the degree-minute-second format. The relatively open exchange of geocoded data makes this recording of the system even more important. The simple rectangular projection, in which latitude and longitude are simply drawn without projection at all, may get the map done, but denies a cartographic tradition going back over 2,000 years and may give incorrect results if used for computations.

4.5.2 The UTM Coordinate System

If we use the ability to georeference other planets as a measure of success, a successful geocoding system for cartography is the Universal Transverse Mercator (UTM) coordinate system. The equatorial Mercator projection, which distorts areas so much at the poles, nevertheless produces minimal distortion along the equator. Lambert modified the Mercator projection into its transverse form in 1772, in which the "equator" instead runs north-south. The effect is to minimize distortion in a narrow strip running from pole to pole.

UTM capitalizes on this fact by dividing the earth up into pole-to-pole strips or zones, each 6 degrees of longitude wide, running from pole to pole. The first zone starts at 180

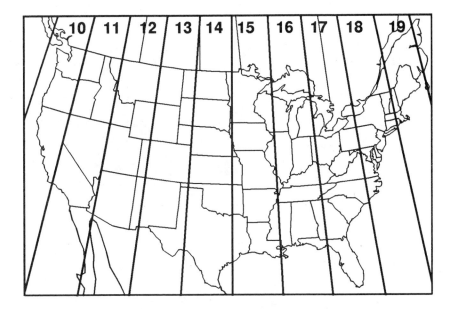

Figure 4.7 Universal Transverse Mercator zones in the 48 contiguous states.

degrees west (or east), at the international date line, and runs east, that is, from 180 degrees west to 174 degrees west. The final zone, zone 60, starts at 174 degrees east and extends east to the date line. The zones therefore increase in number from west to east. For the United States, California falls into zones 10 and 11, while Maine falls into zone 19 (Figure 4.7). Within each zone we draw a transverse Mercator projection centered on the middle of the zone. Thus for zone 1, with longitudes ranging from 180 degrees west to 174 degrees west, the central meridian for the transverse Mercator projection is 177 degrees west. Because the equator meets the central meridian of the system at right angles, we use this point to orient the grid system (Figure 4.8). In reality, the central meridian is sometimes set to a map scale of slightly less than one, making the projection for each zone secant along two lines at true scale parallel to the central meridian.

Two forms of the UTM system are in common use. The first, used for civilian applications, sets up a single grid for each zone. To establish an origin for the zone, we work separately for the two hemispheres. For the southern hemisphere, the zero northing is the South Pole, and we give northings in meters north of this reference point. Fortunately, the meter was originally defined as one–ten millionth of the distance from the pole to the equator, actually measured on the meridian passing through Paris.

Although the distance varies according to which meridian is measured, the value 10 million is sufficient for most cartographic applications. Although the meter has been re-defined in a more precise way, the student may wish to compare the utility of its origin with the origin of the foot, first standardized as one third of the distance from King Henry I's nose to the tip of his fingers.

The numbering of northings starts again at the equator, which is either 10 million meters north in southern hemisphere coordinates or 0 meters north in northern hemi-

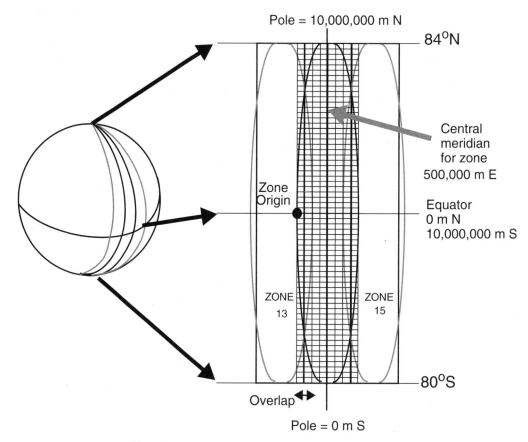

Figure 4.8 The Universal Transverse Mercator coordinate system.

sphere coordinates. Northings then increase to 10 million at the North Pole. Note that as we approach the poles, the distortions of the latitude-longitude grid drift farther and farther from the UTM grid. It is customary, therefore, to use the UTM system neither beyond the land limits of North America, nor for the continent of Antarctica. This means that the limits are 84 degrees north and 80 degrees south. For the polar regions, the Universal Polar Stereographic coordinate system is used.

For eastings a false origin is established beyond the westerly limit of each zone. The actual distance is about half a degree, but the numbering is chosen so that the central meridian has an easting of 500,000 meters. This has the dual advantage of allowing overlap between zones for mapping purposes and of giving all eastings positive numbers. We can tell from our easting if we are east or west of the central meridian, and so the relationship between true north and grid north at any point is known. To give a specific example, Hunter College is located at UTM coordinate 4,513,410 meters north; 587,310 meters east; zone 18, northern hemisphere. This tells us that we are about four-tenths of the way up from the equator to the North Pole, and are west of the central meridian for our zone, which is centered on 75 degrees west of Greenwich. On a map showing Hunter College, UTM grid north would therefore appear to be east of true north.

For geocoding with the UTM system, 16 digits is enough to store the location to a precision of 1 meter, with one digit restricted to a binary (northern or southern hemisphere) and the first digit of the zone restricted to 0 to 6 (60 is the largest zone number). This coordinate system has two real cartographic advantages. First, geometric computations can be performed on geographic data as if they were located not on the surface of a sphere but on a plane. Over small distances, the errors in doing so are minimal, although it should be noted that area computations over large regions are especially cartographically dangerous. Distances and bearings can similarly be computed over small areas.

The second advantage is that the level of precision can be adapted to the application. For many purposes, especially at small scales, the last UTM digit can be dropped, decreasing the resolution to 10 meters. This strategy is often used at scales of 1:250,000 and smaller. Similarly, submeter resolution can be added simply by using decimals in the eastings and northings. In practice, few applications except for precision surveying and geodesy need precision of less than 1 meter, although it is often used to prevent computer rounding error.

4.5.3 The Military Grid Coordinate System

The second form of the UTM coordinate system is the military grid, adopted for use by the U.S. Army (Department of the Army, 1983) and many other organizations. The military grid uses a lettering system to reduce the number of digits needed to isolate a location, since letters can be spelled out using simple words over radio broadcasts. Zones are numbered as before, from 1 to 60 west to east. Within zones, however, 8 degree strips of latitude are lettered from C (80 to 72 degrees south) to X (72 to 84 degrees north: an extended-width strip). The letter designations A, B, Y, and Z are reserved for Universal Polar Stereographic designations. A single rectangle, 6 by 8 degrees, generally falls within about a 1,000 kilometer square on the ground. These grids are referenced by numbers and letters; for example, Hunter College falls into grid cell 18T (Figure 4.9).

Each grid cell is then further subdivided into squares 100,000 meters on a side. Each cell is assigned two additional letter identifiers (Figure 4.10). In the east-west (x) direction, the 100,000-meter squares are lettered starting with A, up to Z, and then repeating around the world, with the exception that the letters I and O are excluded, because they could be confused with numbers. Thus the first column, A, is 100,000 meters wide and starts at 180 degrees west. The alphabet recycles about every 18 degrees and includes about six full-width columns per UTM zone. Several partial columns are given designations nevertheless, so that overlap is possible, and some disappear as the poles are approached.

In the north-south (y) direction, the letters A through V are used (again omitting I and O), starting at the equator and increasing north, and again cycling through the letters as needed. The reverse sequence, starting at V and cycling backwards to A, then back to V, and so forth, is used for the southern hemisphere. Thus a single 100,000-meter grid square can be isolated using a sequence such as 18TWC. Within this area, successively

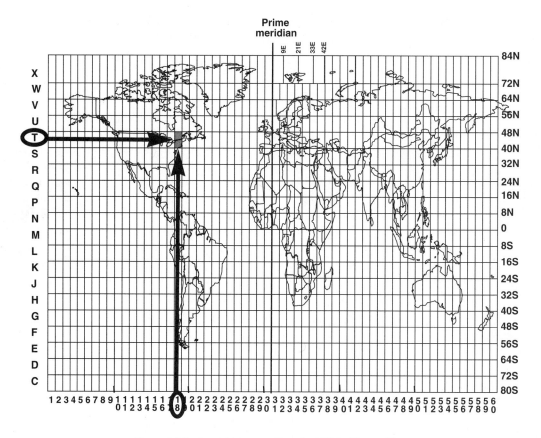

Figure 4.9 Six-by-eight degree cells on the UTM military grid.

accurate locations can be given by more and more pairs of *x* and *y* digits. For example, 18TWC 81 isolates a 10,000-meter square, 18TWC 8713 a 1,000 meter square, and 18TWC 873134 a 100-meter square. These numbers are frequently stored without the global cell designation, especially for small countries or limited areas of interest. Thus WC873134, two letters and six numbers, would give a location to within 100-meter ground accuracy.

The Universal Polar Stereographic (UPS) coordinate system, also part of the military grid, is based on a stereographic map projection centered on each on the poles. The two projections are centered so that the western hemisphere is to the left. For the north polar region, the western zone is designated Y, and the eastern Z, each extending from 84 to 90 degrees north. The prime meridian is used as the right angle for the grid, and 100,000-meter grid cells are simply lettered from A to P running from the bottom (the prime meridian) to the top.

The eastings for the letter designations Y and Z are chosen so that the first column left of the pole is Z and the first to the right is A. For cell Y, R is the first column, while for cell Z, J is the last. For the South Pole, the situation is identical, except that the cells are inverted, that is, the Greenwich meridian is to the top. The cells are designated A to the

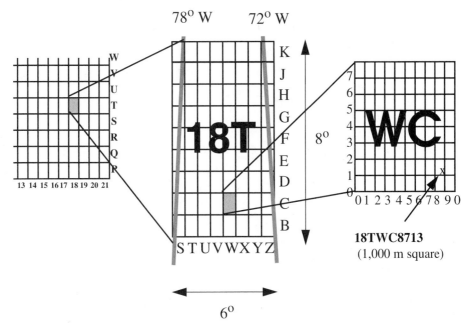

Figure 4.10 Military grid cell letters.

west and B to the east, and since the zone is larger (80 to 90 degrees south), the letters go from A to Z as northings and J to Z and A to R as eastings, respectively (Figure 4.11).

4.5.4 The State Plane Coordinate System

Much georeferencing in the United States uses a system called the State Plane Coordinate System (SPCS). The SPCS is based on feet and has been used for decades to write legal descriptions of properties and engineering projects. Legal documents are probably the least modifiable descriptions on Earth. The SPCS is based upon a different map of each state, except Alaska. States that are elongated north to south, such as California, are drawn on a Lambert Conformal Conic projection. States that are elongated east to west, such as New York, are drawn on a transverse Mercator projection, because the zones are divided into north-south strips. The state is then divided up into zones, the number of which varies from small states, such as Rhode Island with one to as many as five. Some zones have no apparent logic; for example, the state of California has one zone that consists of Los Angeles County alone. Some have more logic, so, for example, Long Island has its own zone for the state of New York. Because there are so many projections to cover the land area, generally the distortion attributable to the map projection is very small, much less than in UTM, where it can approach 1 part in 2,000.

Each zone then has an arbitrarily determined origin that is usually some given number of feet west and south of the southwestern-most point on the map. This again means that

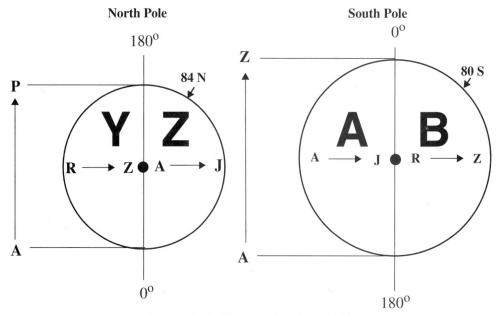

Figure 4.11 100,000 meter cells on the UPS grid.

the eastings and northings all come out as positive numbers. The system then simply gives eastings and northings in feet, often ending up with millions of feet, with no rounding up to miles. The system is slightly more precise than UTM because coordinates are to within a foot rather than a meter, and it can be more accurate over small areas. A disadvantage is the lack of universality. Imagine mapping an area covering the boundary between not only two zones, but two states. This means that you could be working with data that fall into four coordinate systems on two projections. Calculating areas on that basis becomes a set of special case solutions. On the other hand, SPCS is used universally by surveyors all over the United States.

4.5.5 Other Systems

There are, in addition, many other georeferencing systems. Most countries have their own, although many use UTM or the military grid. The National Grid of the United Kingdom uses the lettering system of the Military Grid. In a few cases, particularly Sweden, the national census and other data are directly tied into the coordinates. Within the United States, many private companies and public services use unique systems, usually tied to specific functions such as power lines, or a specific region such as a municipality, or even a single construction project. When using a georeferencing system for geocoding, we should be sure to remain consistent within that system and to record the relationship between the system and latitude and longitude or some other recognized system. Also, we should be sure to use precision and numbers of significant figures that make sense. Can we really measure distances over entire states down to the micrometer or less? And even if we can, is this storage efficient? On the other hand, there is also a tendency to throw away precision needlessly. If we round all cartographic data up to 78 meters, a convenient number determined by an arbitrary orbiting satellite system, how will we deal with newer, higher-resolution systems?

In summary, the universality of the UTM system makes it attractive for world-scale geocoding, while for small areas other systems may be more accurate. Care should be taken to be consistent and accurate, and to use an appropriate level of precision. Fortunately, computer software for mapping allows data to be used from more than one referencing system.

4.6 GEOCODING METHODS

Historically, many different means have been used to geocode. At first, some computer cartographic software actually required maps to be encoded and entered by hand. The hours of monotonous work required for this task made errors common and their correction difficult. Since special purpose digitizing hardware became available, and especially since the cost of this hardware fell substantially, virtually all geocoding has been performed by computer.

4.6.1 Semiautomated Digitizing

There are two distinct technologies for geocoding. The first assists a person in performing the digitizing task; the second performs the task in a completely automated fashion. The first type, *semiautomated digitizing* involves the use of a digitizer or digitizing tablet (Figure 2.2). This technology has developed as computer mapping and computer-aided design have grown and placed new demands on computer hardware.

The digitizing tablet is a digital and electronic equivalent of the drafting table. The major components are a flat surface, onto which a map is usually taped, and a stylus or cursor, which has the capability of signaling to a computer that a point has been selected. As we saw in Chapter 2, the mechanism to capture the location of the point differs. Many systems have connected arms, but most have embedded active wires in the tablet surface which receive an electrical impulse sent by a coil in the cursor. In some rare cases, the cursor transmits a sound, which is picked up and recorded by an array of microphones.

The actual process of digitizing a map proceeds as follows (Figure 4.12). First, the paper map is tailored or preprocessed. If the map is multiple sheets, then the separate sheets should be digitized independently and digitally merged (zipped) later. The next major step, unless annotations have to be made onto the map to assist the geocoding, is to derive a coordinate system for the map. Most applications use UTM, the Military Grid, or latitude and longitude, but many cartographers ignore these standard systems and use hardware coordinates or map inches or millimeters. Map units are sometimes used when precise matching between the digitized map and its source is required. The map is then transformed into geographic coordinates when the editing and proofing is complete.

As a minimum, the coordinate locations of three points are required, usually the upper right easting and northing and the lower left easting and northing, and one other corner. From these points, with their given (user or world) coordinates, and their raw digitizer coordinates, all the parameters can be computed for affine transformations.

This means that the orientation of the map on the tablet need not be exact as is required of some digitizing packages. Many software packages require four of these control points for computing the affine transformation parameters, and it is advisable to repeat digitize points and average coordinates to achieve higher accuracy (see Chapter 9).

The beginning of the digitizing sequence involves selecting the control points and interactively entering their world coordinates. This is a very important step, because an error at this stage would lead to a complex systematic error in every pair of coordinates. After the map is taped to the tablet, it should not be moved without reregistration, and it is preferable to perform the registration only once per map. Ideally, the entire digitizing process should be finished at one sitting, although this is often impossible.

Tape should be placed at each map corner after smoothing the map, and care should be taken to deal with folds and the crinkles that develop during periods of high humidity with certain papers. A stable base product such as Mylar is preferable for digitizing. The lower edge, which will have the cursor and your right sleeve (if you are right-handed) dragged over it many times, should be taped over its entire length. Always permanently record the x and y values of the map control points, ideally digitally and with the geocoded data set. This may allow later recovery of lost resolution or systematic errors.

Digitizing then proceeds with the selection of points. The cursor may have multiple buttons and may be capable of entering text and data without using the keyboard. Voice data entry and commands are also sometimes used. On specialized workstations, there may even be a second tablet with its own mouse or cursor for commands. Errors can be reduced during this process by reading the documentation in advance and by occasionally stopping to review the actual data being generated on the screen.

Points are usually entered one at a time, with a pause after each to enter attributes such as labels or elevations. Lines are entered as strings of points and must be terminated with an end-of-chain signal to determine which point forms the node at the end of the chain. This signal must come from the cursor in some way, either by digitizing a point on a preset menu-area or by hitting a preset key. Unless the chains are to be software-processed for topology, chain-linkage information may need to be entered also, or a point within each polygon digitized and tagged.

Polygons are usually digitized as chains, although sometimes an automatic closure for the last point (snapping) can be performed. Finally, the points should be checked and edited. The digitizing software may contain editing features, such as delete and add a chain or move and snap a node. The software may also support multiple collection modes. *Point mode* simply digitizes one point each time the button on the cursor is pressed. *Stream mode* generates points automatically as the cursor is moved, either one point per unit of time or distance. This mode can easily generate very large data volumes and should be avoided in most cases. Error correction is especially difficult in this mode. *Point select mode* allows switching between point and stream mode. This mode is sometimes used when lines are both geometric and natural, such as when following a straight road and then a river.

Stable base copy of map is prepared by choosing digitizing control points at known locations. Any features to be selected should be marked in advance.

Control points and window limits to be used are marked clearly for use and reuse. Cursor cross hairs should be used at 45 degrees to grid, which should be as fine as possible. Repeat digitizing of controls is advisable for precision.

Control point coordinates are labeled. These should be checked at least twice during tablet setup. Systematic errors will result otherwise. These points can also be used to register test plots.

Map is firmly taped or fixed down to tablet. No movement of the map should be possible. Surface should be flat and free of folds, bubbles, and so forth. Double tape over edges that will be rubbed by elbows and forearms.

Figure 4.12a The semiautomated digitizing process (Part 1).

Control points to be used in the affine transformation are entered one at a time, along with their map coordinates in geocoded coordinate space such as latitude and longitude or UTM meters, or in map inches or millimeters if the affine transformation is to be performed after map editing.

Control point at upper right is entered.

Control point at lower left is entered.

Third control point at lower right is orthogonal, allowing computation of map rotation angle, scaling, and location of tablet versus map original. Map setup is now complete. The setup should be tested by digitizing several points and checking the display and the computed coordinates.

Figure 4.12b The semiautomated digitizing process (Part 2).

Digitizing begins. Map features are traced out using the cursor. Care is taken to capture features accurately, with a suitable level of detail. Points can be selected one at a time or in a stream turned on and off from the cursor. Attributes can be entered as features are completed.

At various stages, and at completion, the map is plotted from software. The plot should be at the exact same scale as the original, allowing overlay. Features should be edited, deleted, or added as appropriate.

The edit plot should be compared with the original, and any necessary final edits should be made.

Final map should be saved carefully. The user can now move ahead to process topology or any other necessary stage before data can be used.

Figure 4.12c The semiautomated digitizing process (Part 3).

At this point the data are ready either for direct integration into the computer mapping software or are ready to be used as input for cartographic transformations to change data structure, map base, or scale. As a human-machine interactive process, the digitizing process has not frequently been studied in detail by cartographers. The process is important to understand, because most of the errors in digitizing can be reduced or eliminated using some simple ergonomic principles (Jenks, 1981).

4.6.2 Automated Digitizing

The second digitizing process is *automated digitizing,* of which there are many types, and, indeed, it is rapidly broadening in scope as a means of data capture. The earliest device for automatic digitizing was the scanner, which receives a sheet map, sometimes clamped to a rotating drum, and which scans the map with very fine increments of distance measuring the radiance, or sometimes transmission of the map when it is illuminated, either with a spot light-source or a laser (Figure 4.13). The finer the resolution, the higher the cost and the larger the data sets. A major difference with this type of digitizing is that lines, features, text, and so forth are scanned at their actual width and must be preprocessed for the computer to recognize specific cartographic objects. Some plotters can double as scanners, and vice versa.

For scanning, maps should be clean and free of folds and marks. Usually, the scanned maps are not the paper products but the film negatives, mylar separations, or the scribed materials that were used in the map production. An alternative scanner is the automatic line follower, a scanner that is manually moved to a line and then left to follow the line automatically. The Altek Apache digitizing cursor is halfway between this and semiautomated digitizing, because it is a manual cursor that has a small scanning window on the cursor itself. Automatic line followers are used primarily for continuous lines, such as contours. These and other scanners are very useful in CADD systems, where input from engineering drawings and sketches is common.

Increasingly, video scanners are becoming important geocoding devices. These scanners are simply television cameras, sometimes with highly enhanced resolution, that can be mounted on a stand and pointed at a stationary image, air photo, or map. Early versions were monochrome only and had limited gray levels. More recently, color versions with many gray levels, and even color look-up tables, have been used. Map separations can be scanned and entered as separate data layers, and data can be sent directly to a microcomputer with local editing, storage, and even image-processing capabilities. Even complex maps such as topographic quadrangles can be scanned by these devices with suitable results, although for a whole quadrangle multiple scans must be used for a reasonable resolution. Again it should be noted that the scanner sees folds, pencil lines, erasures, correction fluid, and coffee stains as easily as cartographic entities. Great care should be taken with scanning not to introduce complex geometric relationships between the map and the image, such as the effects of using different lenses.

Finally, low cost scanners are now available that can read and interpret both documents and text. This is important, because typed and other text can be entered directly from a map and then manipulated and plotted within a mapping system. Simple graphics

Figure 4.13 The automated digitizing process (scanning).
(Photo by the U.S. Geological Survey. Used with permission.)

scanning is rarely adequate for cartographic purposes, but can be used to put a rough sketch into a CADD system for reworking. In this way, first-draft or worksheet sketches can be used as the primary source of information for developing the final map design. Most of the graphics in this book were produced using a combination of scanning maps and photographs and graphics editing using a CADD-like package.

4.7 TOPOLOGICAL GEOCODING

Bearing the goals of geocoding in mind, let's look at a real geocoding method. More important, let's look at how it has evolved over time. The example we use is the geocoding associated with the mapping needs of the U.S. Bureau of the Census, part of the Department of Commerce. This geocoding system is particularly good for seeing how topological geocoding became important over time, and introduces the topic of data structures taken up in the following chapters.

The specific mapping need is to support the decennial census effort (as required by the U.S. Constitution), by generating street-level address maps for use by the thousands of census enumerators. Fairly early on, the use of the computer was recognized as critical. An early system, the address coding guide, was largely a text database, both computerized and manual, that listed all of the street addresses within census enumeration districts.

Included as part of the record was a designation listing the block side of the address, each one given a unique number, as well as geographic information such as census tract, block, ward, and post office codes.

Eighty-eight of the standard metropolitan statistical areas (SMSAs) for the 19th census (1970) were entered manually, with the rest coming from existing computerized mailing lists and post office checks. All 233 SMSAs were covered by this system, with over 40 million addresses.

A need was identified to link this address information with the hand-drawn enumeration maps used in the field. These maps were compiled by asking local communities for all available maps, which were then used as a base for designating the enumeration districts. Understandably, the lack of standardization and variation in accuracy and age of the maps was considerable.

In planning for the 1970 census, a computer-based address coding guide (ACG) was tested in 1967 for New Haven, Connecticut. because a major area of interest in this pilot study was computer mapping, an attempt was made to add geocodes to the ACG files. At that time, digitizing tablets were rare, and the Census Bureau had to design and build its own prototype in-house system. Many maps were digitized using light tables and graph paper. Geocodes with state plane coordinates were added to the ACG records, although some records used latitude and longitude and "map miles." This is one of the earliest cases of using digitally encoded topology to supplement location geocodes. Further use of ACG in this context, however, was abandoned. The process of using the guide had proven too difficult, with too many conflicts, errors and inefficiencies in the digitizing methods used.

The resultant technical steering group, which was overseeing the census use study in New Haven, recommended developing new methods for future censuses. A proposal was made to build geographic base files for the census separately, using graph theory as the underlying concept. Each street, river, railroad track, municipal boundary, or other map feature was considered as a straight line segment. Curved streets were constructed from multiple straight segments. Each node, line segment, and enclosed area was uniquely identifiable within the full network. Line segments were labeled with street names from the base maps, and nodes were numbered sequentially.

The entire system was called dual independent map encoding (DIME), because the basic file was created by computing two independent incidence matrices from the source map, line segment/node, and line segment/enclosed area (Department of Commerce, 1970). DIME, therefore, built substantially upon ACG in that it allowed external checking of the logical consistency of the data by performing topological checks. A DIME file was constructed for New Haven. The major difference between this file and the ACG file was that DIME records contained codes for both sides of a street; that is, the records were segment—rather than block side—based (Figure 4.14). The first few DIME files were used for experimental computer mapping, using such software as MAP01 and SYMAP.

Between the 1970 and 1980 censuses, DIME was updated and extended in scope. The geographic coverage was also extended so that a geographic base file of many SMSAs became available to census users. Automated error-detecting methods were devised that used the topological geocodes to locate and correct errors. Names were standardized and header records were devised for the computer tapes on which the GBF-DIME data sets were distributed. For the 1980 census, virtually all the major SMSAs were covered, with a total of 300,000 enumeration districts. The 1980 GBF-DIME files were used commercially and have remained as important sources of data for a large number of applications.

Analytical flexibility was a key to DIME. Although the geocodes were designed as a way of producing enumeration maps, it was soon realized that DIME was suitable for more general computer mapping. DIME was used for collection of statistical data, automatic generation of centroids, thematic mapping, and automatic address matching, a form of geocoding in itself. Much of this success is because DIME is georeferenced into the census data itself, allowing automatic thematic cartography with thousands of attributes to choose from.

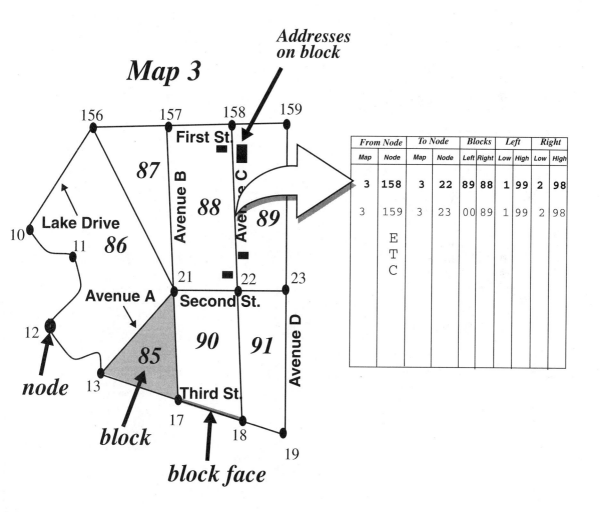

Figure 4.14 Geocoding topologically using the DIME system.

After the use of GBF/DIME for the 1980 census, people started looking at automated thematic mapping as an effective means of understanding the mass of data the census provided. One of the problems for mapping directly from DIME, however, was that most of the segments were street or political boundaries. In addition, they are highly generalized. The demands of detailed computer cartography derived from census data called for a revision of the DIME system. As noted above, an important measure of a geocoding strategy is the ability to adapt to demands previously unforeseen. DIME gave a certain storage efficiency and more analytical flexibility than was at first realized. It produced a lot of labor in geocoding, and there were errors in the original GBF files, but in the long run the system proved invaluable, succeeding in its original goal and going on to different applications.

For the 1990 census, the Census Bureau developed a system called TIGER, for topologically integrated geographic encoding and referencing. TIGER came with a refinement of the DIME terminology (Figure 4.15). Instead of using the block face or street

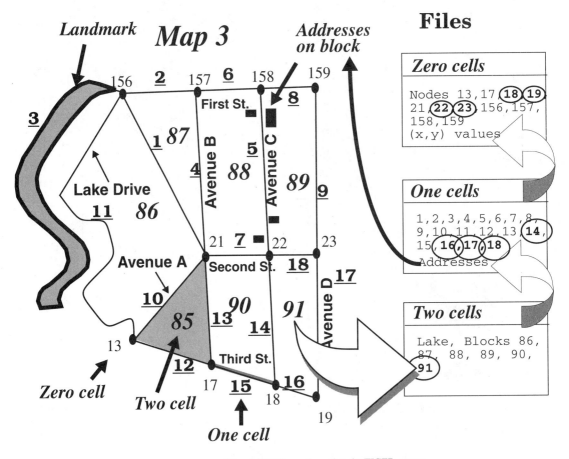

Figure 4.15 Geocoding using the TIGER system.

segment as the basic entity, the TIGER system recognizes cartographic objects of different dimensions. The objects are points (nodes), lines (segments), and areas (blocks, census tracts, or enumeration districts). In the TIGER terminology, points are zero cells, lines are one cells, and areas are two cells.

Under the TIGER system, the digital cartographic boundaries of census tracts follow real things on the earth's surface. Cartographic elements that were digitized included roads, hydrography, political boundaries, railroads, and some miscellaneous features, such as commercial buildings. Each was digitized as a point or a line and assembled to enclose areas. Interactive checking and maintenance software allowed manipulation of these objects, individually and collectively, as well as labeling and consistency checking. The new files were structured topologically and linked together by cross-referencing rather than by having all data in a single file. Thus linkage with the address records is possible, as is the isolation of the data and map base required for a particular thematic map. Choropleth maps, for example, need only the attributes and the two cells, while detailed maps need mostly one cells and labels.

This required a whole new series of maps. A large-scale cooperative effort prepared these maps for the 1990 census. Focal to the effort was having part of the digitizing phase performed in collaboration with the United States Geological Survey. As a result, the Geological Survey was able to convert to a digital basis many of the 1:100,000 series and some of the 1:24,000 series topographic maps.

In the evolution of the use of digital data at the Bureau of the Census, foremost is the collection of massive amounts of digital map data. By changing the basic geocoding method, going from a system based on the block face with a sequential topology to a system based on entities of different dimensions, there has been a substantial increase in analytical flexibility. At the same time, intergovernmental cooperation has reduced the labor in geocoding. The incorporation of topology has allowed a substantial amount of automated error checking, which has improved the quality of the data. The evolution from GBF/DIME to TIGER, therefore, has achieved many of the goals for improving geocoding. In addition, the TIGER data are likely to prove an important archive for future map production of many different types and by many different groups, proving once again that success can be measured by the ability to plan for the unanticipated future applications of geocoding.

4.8 REFERENCES

Chrisman, N. R. (1987). "Efficient Digitizing Through the Combination of Appropriate Hardware and Software for Error Detection." *International Journal of Geographical Information Systems,* vol. 1, no. 3, pp. 265–277.

Department of Commerce (1970). *The DIME Geocoding System*, Report Number 4, Census Use Study. Washington, DC: U.S. Bureau of the Census.

Department of the Army (1973). *Universal Transverse Mercator Grid,* TM 5-241-8, Headquarters, Department of the Army. Washington, DC: U.S. Government Printing Office.

Department of the Army (1983). *Grids and References,* TM 5-241-1, Headquarters, Department of the Army. Washington, DC: U.S. Government Printing Office.

Jenks, G. F. (1981). "Lines, Computers, and Human Frailties." *Annals, Association of American Geographers,* vol. 71, no. 1, pp. 1–10.

Lee, D. and T. Sallee (1970). "A Method of Measuring Shape." *Geographical Review,* vol. 60, pp. 555–563.

Pavlidis, T. (1978). "A Review of Algorithms for Shape Analysis." *Computer Graphics and Image Processing,* vol. 7, pp. 243–258.

Ward, J. R., and M. J. Phillips (1987). "Digitizer Technology: Performance Characteristics and the Effects on the User Interface." *IEEE Computer Graphics and Applications,* April, pp. 31–44.

5

Data Storage and Representation

5.1 STORAGE MEDIA

Computer cartography demands a familiarity with the tools and limitations of digital technology. In Chapter 4 we examined the conversion of map data to digital form using geocoding. When a map consists solely of numbers, the cartographer is faced with many decisions about how to store the map data and how to select methods of representation that allow the data to be used for mapping. In Chapter 2 we surveyed the various types of devices used in computer cartography, especially graphic input and output devices, but we explicitly left out storage devices. The ultimate purpose of geocoding and choosing a representational method and data structure for cartographic data is to store the digital map on some kind of permanent storage device. Ideally, the data should be easy to get back from storage and should also be stored in a way that is both permanent (usable for more than 10 years, at least!) and transportable to other computers. The means of storage is often also the means of distribution. In the days of paper maps, one could purchase the paper over the counter or through the mail, and the information was contained in the printed inks on the paper's surface. Digital map data have to be distributed also. In this chapter we deal with the means of storage, both physical and logical. In Chapter 6 we examine more closely the role of distribution of digital map information.

Over time, improvements in storage technology have drastically reduced the cost and space requirements for storage, and permanence has improved. A key issue for storage remains portability between computers, although distributed computing and networks have relegated this problem to one of archiving. During the initial years of computing, the most widely used storage methods were the punched card and paper tape. These storage methods were not permanent, and they suffered from additional problems, such as sequencing and tearing. Disk and magnetic tape storage replaced cards and paper tape. Fixed or hard disks are usually part of the computer and as such are not transportable, unlike their smaller but portable equivalent, the removable disk. Removable disks, however, are fragile and bulky and can be damaged by strong magnetic fields.

Floppy disks, introduced when microcomputers became available, are relatively rigid, so a preferable term in use is *diskette*. Older disks were enclosed in a plastic envelope containing the thin magnetic sheet that holds the data. Early types were 133 mm (5.25 inches) and 203 mm (8.0 inches) in diameter, but they are rapidly being replaced by 89-mm (3.5-inch) disks, which have a more rigid plastic case and can support higher storage densities. Such a disk has been used to distribute the software and data for use with this book, and is enclosed in a plastic pocket at the back of the book.

Probably the longest surviving and most reliable storage medium is magnetic tape, which comes both as loose reels and tape cartridges. Different lengths of tape and different numbers of bits per inch (BPI) of storage density allow different amounts of data to be stored. The cost, reliability, transferability, and high density of magnetic tape made this the normal medium for data and software distribution for large computers. The advent of workstations has led to several new tape formats, including the 4-mm tape cartridge and the 8-mm tape. The latter are highly flexible and extremely compact and are capable of storing between 2.3 and 5.0 gigabytes of data, depending on blocking and structure, at a cost of only a few dollars. This change clearly indicates that basic level digital cartography now faces few physical limitations of storage volume, even on microcomputers.

More recently, mass storage has become possible using optical disk technology. Many optical disks are read-only and can play all or part of a record but cannot record. The first generation of optical disks used compact discs as read-only memory, and disk templates had to be written in much the same way that a book is printed. More recent optical media allow both reading and writing. The advantage of optical media is the size of the storage, sometimes approaching gigabytes of storage even for a microcomputer. Almost all topographic map coverage for a single state, both current and historical, would fit onto a single optical disk. Several companies now distribute atlases, gazetteers, and even street maps in CD-ROM, such as the Map Explorer software and data by DeLorme. The disks themselves are comparatively fragile and require special holders for placement into a drive. Nevertheless, the prices are now so low that they are a favored means of data and software distribution. The per item costs for disks that are sold in volume can approach the cost of cartridge tape. For example, many government agencies distribute CD-ROM data sets for about $30.

Recently, optical disk "juke-boxes" have become available. These devices allow the user to select, through software, which CD-ROM is to be loaded onto the drive to be read. Stacks of CD-ROMs are therefore available to the user, making hundreds of gigabytes accessible. So large have storage volumes now become that discussions use the terms *terabyte* (1,000 gigabytes) and even *petabyte* (1,000,000 gigabyte). The archive of Landsat data at the EROS Data Center in Sioux Falls, South Dakota, for example, now contains about 40 terabytes or 0.04 petabytes. Perhaps a petabyte of storage for a desktop cartographic workstation may be possible, perhaps sooner than we think! Such a situation ensures a vast redundancy in the supply of digital map data, guaranteeing its survival in the same way as publishing ensures the survival of the information content of a book. Similarly, every cartographer may eventually have access to all the maps and images ever created.

Even reliable storage such as disk memory has to be backed up to a more permanent storage medium to guard against failure and loss. Backups are either full-system or incremental, in which every new or changed file on the entire system is copied. All files are copied onto magnetic tape, and the tapes are archived. If anybody inadvertently deletes a file, or if something goes wrong with the computer, data can be retrieved as they were saved on a given day. Magnetic tape readers are available for microcomputers and are usually used to back up a hard disk on magnetic tape. Even microcomputer hard disks must occasionally be backed up to diskette or tape, which are more likely to survive catastrophic events than hard disks.

5.2 INTERNAL REPRESENTATION

The various storage media in use are simply the physical mechanisms for storing digital cartographic (and other) data. To gain insight into computer cartography, we still have to understand how it is that the data are translated into the physical storage properties of the particular medium, be it a magnetic tape, an optical disk, or a diskette. A basic problem is that although a programmer or a cartographer may seek to access data at random, or by a spatial property, nevertheless on the physical storage device the data are stretched out sequentially, one after the other. Using any storage medium involves keeping track of information about where different data are stored on that disk, or tape, or whatever the medium used. This involves blocking, or dividing the data into manageable chunks, and requires a mechanism for dividing the disk into blocks and sectors. This blocking is independent of any spatial data structure that we may impose upon the digital cartographic data, and even of the file structure used to structure data.

A diskette has radial sectors broken up into blocks or areas reserved for storage of different kinds. The operating system is designed to keep track of this information for you. The link between the actual "map" of the blocks, the sectors, and the physical disk itself is the directory. The *directory* is an area of the disk reserved for information about where other things are on the disk. All operating systems have a command to allow you to see what files occupy your storage, although most mask from the user the specifics about where the storage is actually located. Many computer operating systems allow you to have access to much more storage of one kind, such as RAM, than is apparently available by connecting memory of another type, such as disk for RAM. This method is called virtual memory.

Addresses are locations in storage, both permanent and temporary, where information can be stored. The larger the computer's memory, the larger the range of addresses available for storage of information. The very lowest level of address normally contains the operating system. The basic storage unit is the lowest-level piece of information we can fit into or retrieve from an address. On an optical storage system, addresses relate to resolution. On a mechanical or electrical system, addresses are related to the ability to hold a magnetic or electrical charge. Storage devices are state storage mechanisms; they can store on or off, binary 1 or 0, or one bit. A capacitor can hold a charge or not. An optical reader may see a bar code with a black line or a white line.

The mathematics that corresponds to this bit logic is called *Boolean mathematics,* and it works in number base two, the counting mechanism we would use if we had only one finger on each hand. Counting is in the sequence of zero, one. As we string the bits to-gether we can build bigger and bigger numbers. There is a clear correspondence between distribution of digits in binary numbers and what the memory storage actually looks like. On the older storage media, such as cards and paper tape, you could actually see the bi-nary representations as rows of holes punched out of the paper. On magnetic media, the tape either holds a charge or does not.

As an example, the binary number 1010 0011 0001 corresponds to the decimal num-ber 2,609. Notice that just as we often break off the 2 from the 609 with a comma for ease of interpretation (most other nations use a space, an apostrophe, or a period), we use spac-es between groups of four binary digits or bits. This has the distinct advantage that it is easy to represent the number in other number bases. The two most common alternative number bases are base eight (octal) and base sixteen (hexadecimal). This is because one octal digit corresponds directly to three bits, while one hexadecimal digit corresponds di-rectly to four bits.

By far the most used is hexadecimal, in which the counting numbers are 0, 1, 2, 3, 4, 5, 6, 7, 8 ,9 , A, B, C, D, E, and F. Each cluster of four bits can be represented as a single hexadecimal digit. Thus 1010 translates to decimal 10 or hexadecimal A; 0011 translates to decimal 3 or hexadecimal 3; and 0001 translates to decimal 1 and hexadecimal 1 (Fig-ure 5.1). So the binary number above can be represented as the three hexadecimal digits A31.

To write the binary equivalent of this value onto the storage medium, many different systems are used. Some write the bit values left to right, some right to left, and some in-vert the value of the bits byte by byte, a process known as byte swapping. The differences relate to whether the least significant bit (that is, right-most logically) is stored first or last. A cluster of four bits can be represented by a single hexadecimal digit. Two hexa-decimal digits together are called a byte. We use the term *word* to refer to the grouping of bytes which the computer itself uses as the basis for storage and computation. Thus we

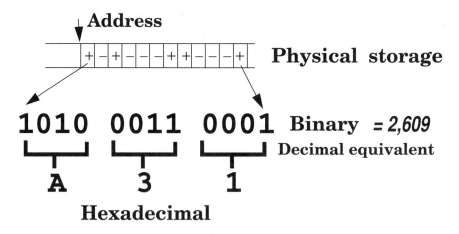

Figure 5.1 Decimal, binary and hexadecimal representation of the number 2,609.

may hear that certain chips are 16-bit or that a particular computer has a 32-bit word. In the early days of microcomputers, there was a big difference in the size of a word on a microcomputer and the size of a word on a large computer. There were even differences company by company on what the size of a word was on a large computer. Today, even some small computers use 32-bit words, while the very largest and most powerful computers have moved on to 64-bit words or other sizes. A larger word size allows a greater range of addressing and more precision on computations.

Because hexadecimal is an intermediate internal representation of the data and programs that are stored on a computer, many computer users never need know this level of detail. Similarly, we may be able to drive a car without opening the hood. When the details under the hood become important is when something is wrong with the car, when we decide to build a new car or to improve the one we have. Programmers make frequent use of binary and hexadecimal, but as a computer user the only time you may see hexadecimal is when something goes wrong. When a program or operating system crashes, it sometimes tries to send a message to assist in finding what went wrong. What the computer sends in the message is an address, or a location in memory, where the error occurred, and usually it gives the address in hexadecimal.

Some scientific pocket calculators do binary, octal, and hexadecimal arithmetic. Using these calculators is a good way to get some experience in using these number systems, although using them for shopping or balancing a checkbook can be confusing. Number operations combine binary digits with operators called AND, OR, and NOT and their inverse combinations (NAND and NOR). AND simply matches two binary numbers and results in a 1 only if both bits are 1. OR results in a 1 if either bit is 1. NOT simply results in the opposite of a bit, zero replaces one and vice versa. Combinations of these operations between numbers stored in memory locations—called registers—are how the computer performs arithmetic.

Hexadecimal codes can be used to number the available storage locations in memory, which are inherently sequential, so that they can be matched with their data contents. Having found the storage location, however, we may wish to retrieve what we find there or change the memory to what we want. If the memory locations themselves held hexadecimal digits, then the computer would only be able to store numbers, and only positive numbers at that.

To get over this limitation, we have developed a standard way of encoding the more complex sequences that we actually use, numbers, characters (such as upper- and lowercase letters), and special characters (such as brackets and exclamation points). We do not actually store the letter *e*, or the number *0*, we store a representation of them in these basic storage codes. What gets stored in the memory locations pointed to by hexadecimal addresses are binary strings that correspond to character sets.

Generally, the most commonly accepted character set is the ASCII (American Standard Code for Information Interchange) set. The ASCII character set stores one character to one byte, giving us 255 possible codes, although only 127 are normally used (that is, 7 of the 8 bits). The first code is called null, or nothing (*not* zero). The next codes correspond to `control` a through z. Remember that on a computer the `control` key is just like "shift" on a typewriter, the difference between uppercase a (A) and lowercase a (a) is to hold down the shift and hit a, or do not hold down the shift and hit a. `Control` works in exactly the same way as `shift`. Just as we have upper- and lowercase characters, we have control characters in the code. ASCII control codes go from hexadecimal 01, which is `control-a`, through hexadecimal 1A, which is `control-z`. ASCII code 1B corresponds to the `escape` key. Many escape sequences do special things, such as graphics. When they get ASCII code 1B hexadecimal (27 decimal), many graphics devices interpret the next piece of data as a command rather than data. This explains why listing a binary file on a monitor often produces unexpected results!

ASCII code hexadecimal 20 (decimal 32) is a space; then we go through all the special characters, exclamation points, pound signs, periods, and so forth. At ASCII code hexadecimal 30 (decimal 48) we start with numeral zero and go up through ASCII code hexadecimal 39 (decimal 57), which is numeral nine. At hexadecimal 41 (decimal 65) we start with uppercase a, up to hexadecimal 5A (decimal 90), which is uppercase z. At ASCII hexadecimal 61 (decimal 97) we start with lowercase a, and at hexadecimal 7A (decimal 122), we are at lowercase z.

Internally, therefore, what actually get stored on the storage medium used for computer cartography are the binary representations of ASCII codes. These values are written into hexadecimally referenced locations in either temporary or permanent memory. The advantages of storing ASCII codes are many and include being able to view and edit the data, as well as maintain a high degree of portability between computers. As we will see, however, using ASCII codes often leads to memory management problems, and these codes are not optimal for storing the raw numbers into which digital maps have been converted by geocoding.

5.3 STORAGE EFFICIENCY AND DATA COMPRESSION

An important goal for the storage of data within computers is to achieve storage efficiency. We often have to make a trade-off between storage requirements and the ease of use of cartographic data. Ease of use is reflected by several competing goals: achieving rapid access to data, sustaining fidelity, allowing concurrent access to multiple users, and facilitating data update and maintenance. In Chapter 4 we noted that one of the fundamental characteristics of geographic and cartographic data is sheer volume.

Geographic data sets are typically very large. This means that storage efficiency problems are pertinent to handling cartographic data. Frequently, we are concerned with how we can reconfigure the data to fit into our particular storage device or to fit our logic or reasoning system. To be efficient with storage, we seek to retain as much *cartographic information* with as little storage as possible.

Storage efficiency concerns two sets of storage demands. First, the entire digital cartographic data set must reside in permanent storage at least somewhere on a network and be accessible to the mapping program as required. Second, a computer cartographic program must bring part of or all the map data into the random access memory (RAM) of the computer for display, analysis, and editing. The computer program moves the data from storage into a data structure in RAM supported within the program itself. Data in RAM are available much more rapidly to the program. Data in secondary storage must be brought into and out of the RAM, a process known as I/O, or input and output. To move data between these two types of storage is time consuming. Because digital cartographic databases are typically large, I/O is the most time-consuming part of producing maps by computer. When only part of the map can fit into the RAM at any given time, such as on a microcomputer that can only access RAM in 640 kilobyte partitions, memory management can become a significant part of the computer mapping software.

The storage efficiency of cartographic data in RAM is determined by the precision of the data and the suitability of the programming data structure. The file size and byte-by-byte character of the data are determined by the physical data structure, (that is, how the logical spatial constructs map onto the computer's memory). The physical structure is the "translator" of the logical structure, the theoretical basis of how the data structure encodes space and spatial relationships, into digits.

A programming data structure is made up of the physical structure into which the data have been put, and the power of the logical data structure as a construct for solving the particular mapping purpose. Obviously, when an effective logical structure is matched by an efficient physical structure, both the analytical and the computer cartographer benefit. The distinction between logical and physical data structures is important and allows different degrees of storage efficiency to be attained.

In Chapter 4 we considered the merits of various graphic input devices. We noted that there are two major types, semiautomatic and automatic, and that historically these structures have influenced programming data structures and storage. In general, the semi-automatic devices produce vector data and the automatic produce raster data. Vector data can be very storage efficient, because we only need to capture information when there is information worth recording.

For example, on the right map projection we can represent the boundary of the state of Colorado with only four points. Curved lines are sometimes a little less efficient because to maximize the amount of information stored we have to concentrate data in areas where the boundary changes in direction. In digitizing a line we capture more points in areas of extreme curvature than in areas that are very straight. This introduces the concept of *information content*. What is a high information content point and what is a low information content point?

Figure 5.2 shows some lines with different degrees of information content at different points. On a straight line, the two end points are very high information points, because they contain information about where the line begins and ends, without which it is impossible to draw the line. A point along the straight line is redundant, because it adds almost nothing to the information content. It does, however, add to the data volume. To add a third x and y value along the straight segment, we have to increase the x and y data storage volume by a third, but we get no corresponding increase in the information content. To achieve storage efficiency, we seek to minimize the redundancy within the data and to maximize the information content. In a vector system we can vary the sampling density to correspond to what we are mapping.

This concept does not apply only to lines, but to points and areas, too. For example, on a surficial geology map, for which we know that there are two basic geology types with one twice as abundant as the other, we would anticipate two different categories for the map area, say schist and gneiss. We know that there is a boundary that runs somewhere within the map area. How would we sample that area to achieve the maximum amount of information content in the data versus the minimum amount of redundancy?

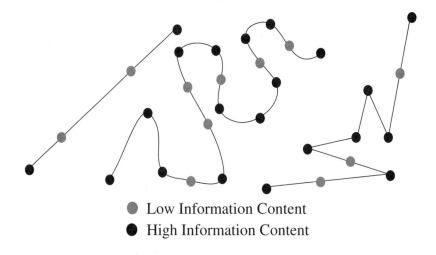

● Low Information Content
● High Information Content

Figure 5.2 Lines with points of different information content

It would be pointless (or rather point-full!) to put a grid over the map and take soil samples at the intersection of every grid point. Most of our data would be redundant , and the sampling strategy would probably miss the only significant feature, the boundary. If we know there is a boundary, we need only a few samples over the map area but many along the boundary itself. This would locate the boundary on the map. Any other sampling strategy would contain redundancy. A vector-based system, in this case irregularly distributed points rather than lines, allows us then to seek out the highest information content areas on the map and to concentrate our sampling effort on those. This strategy, however, assumes some prior knowledge about the nature of the geographic distribution involved.

Raster systems work differently. Using a grid, there is one data value for every grid cell, or grid intersection point per grid. This means that we have a large amount of embedded redundancy and high data volumes for the same geographic area. This is the same as uniform sampling. Within a lake, for example, all grid points simply repeat that this is still a lake.

This strategy is most suited to geographic phenomena that change continuously, such as elevation and physical properties, rather than discontinuous properties, such as surface geology and human characteristics such as rural land-use. Even so, a grid resolution suitable for one geographic region or for a particular phenomenon with distinct geographic properties is usually unsuitable for another. Thus we would use a different grid spacing for elevations on the Great Plains than for elevations on the Rocky Mountains.

Grids happen, however, to be a convenient measurement mechanism. Thus grids are often used for large areas at small scales, while vectors are frequently used for small areas at large scales. The two data structures are really answering two different questions and making two different demands on the data. The two systems are both space-sampling strategies, one regular and the other irregular; one with constant resolution and one with variable resolution. If we accept a direct relationship between scale and resolution, then vector and raster conversions are a problem of scale transformation. Tobler has suggested the single term *resel,* or resolution element, to apply to both regular and irregular areal sampling distributions, with the *pixel,* a square or rectangular resel, as a special case.

5.3.1 Storing Coordinates

A distinct link between logical and physical data storage is important for storage efficiency. In this chapter we consider the physical aspects of storage. In Chapter 6, we consider the major logical map data structures. We saw in Chapter 4 that under the UTM coordinate system we could represent the location of almost any point on earth to a precision of 1 meter by using a string such as

4,513,410 m N; 587,310 m E; Zone 18, N

During the digitizing stage of geocoding, at some stage we moved a cursor to this point and pressed a button. What geocode does this process transfer from the map to storage? The full reference consists of 15 digits, plus a binary value for the hemisphere (north or south). The whole string, as shown above, is actually 32 characters, 15 digits, four commas, one blank, two semicolons, nine letters, and an end-of-line character to separate this from the next line. Simply storing that the values consist of a UTM reference for zone 18 N as a northing, followed by an easting, reduces the storage to 15 characters

4513410 587310

So by storing a line such as

UTM Zone 18 North, Northing then Easting

only once, for 41 bytes (one ASCII code per character, plus the new-line), we can save 20 bytes per record. For 100,000 records, we save 2 megabytes at a cost of 41 bytes. Let's see if we can do better than that. Zone numbers go from 1 to 60.

When we are designing storage systems, we start by looking at the maximum value we need to fit in any field. We only need to store zone numbers between 1 and 60. For the hemisphere we can store either 1 or 0, one binary digit, instead of a whole byte. What is the maximum easting? If we reached 1,000,000 we would be in the overlap between zones, so for all practical purposes 999,999 is the maximum.

The maximum northing is 10 million. Because 10,000,000 south can be represented as 0 north, and because we cut off at 84 degrees north, the maximum northing is 9,999,999. Thus in reality each record can consist of two numbers, one between 0 and 9,999,999 and one between 0 and 999,999. Using hexadecimal, so that digits correspond to half-bytes, these numbers are 98 96 7F and 0F 42 3F, requiring a total of only six bytes per coordinate pair. Furthermore, if these bytes are written sequentially, that is, without the new-line characters, our 100,000 records need occupy only 600 kilobytes. The data set, which formerly would have occupied a significant portion of a hard disk, would now fit easily into RAM on a large microcomputer.

If we took the additional step of dropping the last digit, i.e., giving our data a precision of 10 meters and storing the first 100-km-grid-square letter and number designation as shown in Chapter 4, we need store only eight digits, four each for x and y, making only 10 ASCII characters (bytes) per record, or two values with a maximum of hexadecimal 27 0F, making four bytes per record.

The 100,000 records now take up 1 megabyte as ASCII codes or 400 kilobytes as binary sequential. By simply economizing on the storage representation of a record we have reduced a data file from 3.5 megabytes to 400 kilobytes.

This is an example of achieving storage efficiency by physical data compression, reducing the actual number of bytes required to store information. The storage improvements achieved here, a compression to 11.4% of the (rather padded) original size, were achieved in two ways. First, we used physical compression, achieving savings mostly by using binary instead of ASCII codes. This compression is directly reversible.

Second, we dropped the last digit of the data, saving space but making the compression a noninvertible transformation of the data, actually a scale transformation since we have spatially generalized the data. Dropping the last digit may also create duplicate data points, which can be reduced to a single instance. It is impossible to get this detail back accurately, so a scale change is partly a physical but also partly a logical compression of data storage volume.

The example above is based on reality. The U.S. Geological Survey's land-use and land-cover (LULC) data use a similar compression method. The LULC data in a format known as GIRAS (geographic information retrieval and analysis system) are the digital equivalents of the 1:250,000 land-use and land-cover maps, available for the whole country. In these data sets, which are also topologically encoded, eastings and northings are relative to the nearest 100,000 meter UTM intersection to the west and south of the map area.

This is equivalent to storing once the letter and grid number designations from the Military Grid, covered in Chapter 4. These are stored separately, as headers or "metadata" (data about data) once per file. The eastings and northings are then only five digits long. In addition, the minimum resolution is set to 10 meters rather than 1 meter; in other words, they lop off the last digit. In storing an arbitrary origin once, the zone number and the hemisphere are stored also. This leaves eight bytes for each point because the files are ASCII. No new-line characters are necessary because the points file is simply written sequentially or in some data structure.

An effective means of achieving optimum storage is to limit the geographic area of coverage. Examples are counties, cities, municipalities, and states. Even within these areas, data use can be optimized by storing data to reflect use, such as by making the most frequently used data the easiest to retrieve from storage.

This is common when data bases are tiled, or divided into squares or rectangles across the area of interest. Another way is to make the data structured by scale and purpose, so that where low-resolution data are suitable for mapping needs, they are used rather than the entire data set.

5.3.2 Compressing Coordinate Storage

Some digital cartographic databases or mapping systems have differential accuracy levels embedded in them. In these systems, for each x and y value there is another value, an identifier, that says "Include this point at this level." To make a very highly generalized map, we would scan this last character and include the point only if it meets the lowest generalization level, such as 9, which would be the least detailed. A level 9 map would simply exclude all points with level less than 9. A level 1 map would include every single point. Using this method, we can embed different levels of resolution within the same data set, at the cost of storing one more character per point. An effective way to determine the appropriate resolution factor is to apply a line generalization method, such as the Douglas-Peucker technique, discussed in Chapter 10. Generalizing maps can significantly reduce the amount of storage required for them.

Many data compression methods are suitable for all data files, whether spatial or not, whereas others are unique to spatial data and even a particular data structure such as a grid. Many data compression schemes get much of their saving by converting from ASCII into alternative representations. A popular method of physical data compression is Huffman coding. Huffman coding puts an entire file through a one-step process, the result of which can be a significant saving in storage. The process is entirely reversible, and in doing so the reconstructed file is exactly the same as the original. Files with much redundancy in them, many blank lines and columns for example, or large numbers of similar digits compact significantly with Huffman coding. This is often the case with map coordinate files. Huffman coding works on ASCII files and performs a frequency analysis of the ASCII codes. The ASCII codes are then replaced by a new set of Huffman codes of variable length, with the very shortest length codes corresponding to the most frequently occurring ASCII codes (Held, 1983).

Huffman and other compression schemes are often made available to users as part of the operating system or as utilities. MS-DOS 6.0, for example, implements compression of files to approximately double the apparent size of a hard disk. File transfer by diskette often takes advantage of file compression using public domain utilities such as ARC and PKZIP, which can link files together and compress them, allowing a limited-size disk to transfer large files. The Unix operating system contains the utility compress, which creates a file with a .z extension of considerably reduced size. Compressed files are particularly useful for transfer of files over networks, where the highest information content per byte during the transfer is desirable. Similarly, the GIF and JPEG file formats use compression schemes for storing images (Kay and Levine, 1992).

There is necessarily a penalty for achieving maximum storage efficiency, and what we usually sacrifice is ease of retrieval, or analytical flexibility. Especially now that the cost and limitations of data storage are falling, the savings of space may not be worth the loss of flexibility or precision used to achieve storage efficiency. As a rule, ASCII data representation is superior for cartographic data unless the mapping is in a production environment, because errors can be detected simply by looking at the data with an editor, even for very large data sets and very rapid browsing.

The human eye and brain are powerful error detectors even for large numbers of records. The most effective error detection (and correction) strategy is when the eye/brain method is coupled with automatic error-detection routines. Although data compression may be desirable for data transfer through networks, or archiving on tapes, cartographic data are best represented as the actual location coordinates they ultimately must consist of.

Given the basic means by which data can be stored and compressed on computer storage media, which methods are used in computer cartography, and why? In Section 5.4 we examine some of the formats that have been developed and used, especially by government agencies, and determine the significance of the spatial data transfer standard.

5.4 DATA STORAGE FORMATS FOR CARTOGRAPHY

The beauty, and also the problem, of standards is that there are so many to choose from! Because most digital cartographic data archives were developed piecemeal over a long period and for an enormous number of applications, there are a great many proposed standard formats. The need for standardization is most apparent when data must be transported between applications and between computer systems. The most successful data formats are those that have withstood transportation between systems and those that have had the most use and documentation. In this section, we examine in detail the formats used in the past by the major producers of digital cartographic data within the U.S. federal government—the Defense Mapping Agency (DMA), the National Ocean Service (NOS), the United States Geological Survey (USGS)—plus a widely distributed data format, the World Data Bank. The logic of the Bureau of the Census's formats was discussed in Section 4.7. Although data formats related to specific file and byte structures have often become *de facto* standards, they are rarely standard in the more formal sense. In Chapter 6 we examine the spatial data transfer and other formal data standards and how they relate to the future provision of digital map data.

5.4.1 Formats at the U.S. Geological Survey

Digital cartographic data from the U.S. Geological Survey are distributed by the Earth Science Information Centers as part of the National Mapping Program. USGS digital data fall into four categories: *digital line graphs* (DLGs), *digital elevation models* (DEMs), *land-use and land-cover digital data* (GIRAS), and *digital cartographic text* (Geographic Names Information System, GNIS). The USGS continues to improve coverage of the United States and distributes the map data products on computer tape, floppy disk, and CD-ROM.

The DLG data are digital equivalents of the 7.5- and 15-minute USGS sheet maps, as well as the more generalized 1:100,000 maps and the National Atlas regional-level 1:2,000,000 maps. The DEM data are land surface elevations, at both 1:24,000 with a ground spacing of 30 meters using the UTM grid, and 1:250,000 with a ground spacing of 3-arc seconds, the same as the DTED data described later. The land-use and land-cover data are digital versions of the 1:100,000, and 1:250,000 land-use and land-cover and associated maps (the interim land-cover mapping program for Alaska). Finally, the GNIS is a unique data set that includes the text with its attributes from a large number of USGS map products. By 1994, about 15% of the United States was completely digitized from the 1:24,000 maps, 50% of the 7.5 minute DEMS were finished, and the 1:100,000, GIRAS, and 3-arc second DEMs were entirely finished.

As their name suggests, the digital line graphs are vector encoded and divided by map source scale as separate data sets; the scales are 1:24,000, 1:100,000 and 1:2,000,000. At the 1:24,000 scale, the data consist of boundaries (political and administrative, such as the municipalities, federal lands, and national parks), hydrography (standing water, flowing water, and wetlands), the boundaries of the public land survey system (down to sections in the township and range system), transportation (including roads, trails, railroads, pipelines, and transmission lines), and other significant fabricated structures, such as built-up areas and shopping centers. These maps were digitized from 7.5- or 15-minute quadrangles, or from their compilation products. The 1:100,000 data sets are similarly structured and are subdivided by groups of files covering 30- by 30-minute blocks, from the 1:100,000 scale topographic maps. The 1:2,000,000 maps, digitized from the 1970 National Atlas of the United States of America, follow the same DLG format. Twenty-one digital map sheets cover the United States, fifteen for the coterminous states, five for Alaska, and one for Hawaii. Coverage for New York, for example, includes all the northeastern states, from New York to Maine.

Although these data were originally to consist of three levels of accuracy and coding, in fact all data are provided to the maximum level, that of DLG-3. DLG-3 data are topologically structured and consist of nodes, lines, and areas structured in a manner that explicitly expresses logical geometric relationships. The coded geographic properties are adjacency and connectivity. This allows both plotting of the data for computer cartography and the use of derived information in analytical cartography. The coordinate system is local to the map in thousandths of an inch, but parameters for conversion to UTM are provided in the header files. The latter fact shows clearly how these data relate closely to their sources as paper map products.

The basic cartographic objects in the files are nodes and lines, with areas consisting of labels and links to the other objects. Lines begin and end at a node and are geocoded with the left- and right-hand areas they divide. Lines connect at nodes, and no line crosses itself or another line. Islands are lines that connect at a dummy node; they are termed degenerate lines. Areas are delimited by lines and contain a point, to which is assigned a label for the area. In addition, lines have attribute codes that determine what cartographic entity is being represented. These codes are based on the USGS 7.5-minute map series symbol list. Figure 5.3 shows a typical segment from a DLG and gives examples of different attributes.

An ambitious attempt is now under way to revise the DLG format substantially to encode the spatial relationships between the geographic features on the ground more fully, rather than between their cartographic representations on the map. The new data format, known as DLG-E (Enhanced) will eventually replace the older DLG format. The DLG-E format closely reflects the terminology and data model of the Spatial Data Transfer Standards.

The model is based on features, and includes a set of feature codes for objects such as bridges and rivers as well as a set of topological relationships. As such, DLG-E is more of a move toward the DMAs approach of using integrated topology in map encoding.

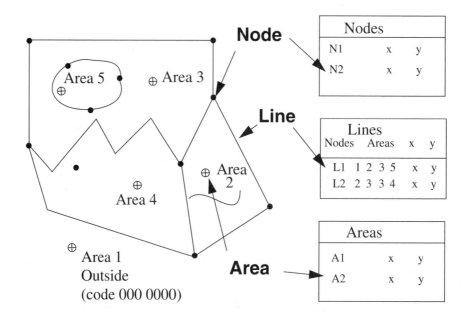

Figure 5.3 Sample digital line graph coding format

Most interesting is the inclusion of text, such as place names, as attributes of features, a significant and important diversion from the current format, in which text is encoded in a separate and unrelated file called GNIS (Geographic Names Information System). A complete discussion of the DLG-E format is contained in Guptill (1990).

The Digital Elevation Models are digitally encoded sample measurements of the surface elevation of topography. Two source map scales are involved, 1:24,000 and 1:250,000, and there are four different map generation technologies. The 1:250,000 data are 3-arc second increments of latitude and longitude for 1-degree blocks covering most of the United States. These data are very similar to the DMA's DTED data and in fact are derived from them. Each 1-degree block contains 1,201 elevation profiles, with 1,201 samples in each profile. Elevations are in meters relative to mean sea level, and latitude/longitude uses the 1972 World Geodetic System datum. Figure 5.4 shows how this data set is logically organized. These sets are distributed on computer tape, with documented headers and formats.

The 1:24,000 data correspond to the 7.5-minute quadrangles and are in UTM coordinates. As such, the boundaries of the map are not square, and the number of elevation samples varies by south-to-north profile over the map. Each profile has a local datum, given in a profile header at the start of each string of samples. The files are ASCII records, stored sequentially on magnetic tape. Figure 5.5 shows this logical format, and it is important to note that the physical structure depends somewhat on which side of the central meridian for the UTM zone the data set falls.

The last USGS data set to be considered here is the digital land-use and land-cover data. Although now largely out of date, these data sets are available by 1-degree blocks

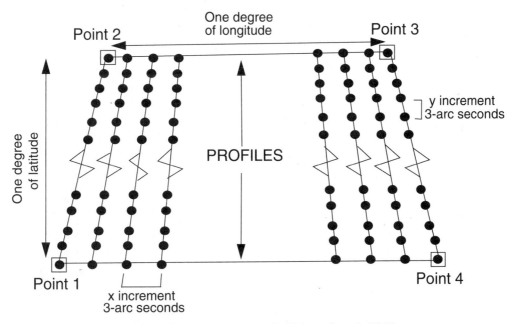

Figure 5.4 1:250,000 3-arc second DEM format from the USGS.

for the entire country and were digitized from the land-use and land-cover sheet maps, with source dates going back to the late 1970s. The program was completed in the early 1980s, leaving an important baseline data set for the study of land-cover change.

Each 1-degree block is split into sections to make files of manageable size; for example, the Albany, New York, 1-degree quadrangle has 15 sections. Data within blocks are stored in GIRAS format. Because the maps consist simply of land-cover polygons with their attributes, most of the data are nodes, lines, and links between the lines to create polygons—a loose topological structure. The map and section headers are followed by an arc records file, which consists of pointers (sequential links) into the next file, which contains the actual coordinates. The coordinates are in UTM and are truncated by rounding to the nearest 10 meters and by using the nearest 100,000 grid intersection.

Many of the USGS data sets are now available directly through the Internet by `ftp`. Chapter 6 gives details about how to search for these files. They are maintained on USGS servers usually listed alphabetically by the name of the map quadrangle concerned. This means that to retrieve the data set, the cartographer should both know what data are needed and what format the data are to be found in. Many major mapping and GIS software packages can now read these files directly, relieving the cartographer of the need to write computer programs to deal with the data formats.

The USGS also now distributes data on land-cover derived from classifications of NOAA's AVHRR (Advanced Very High Resolution Radiometer) measurements. These data are distributed by the EROS Data Center on CD-ROM, with a ground resolution of one kilometer. Biweekly composites showing a vegetation index for North America are available, and efforts are under way to release a global AVHRR data set at this resolution.

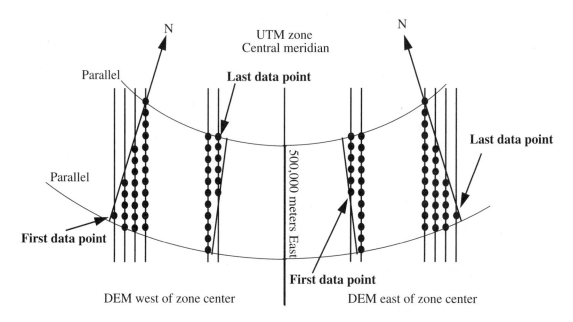

Figure 5.5 1:24,000 30 meter DEM format from the USGS.

5.4.2 The World Data Banks and the Digital Chart of the World

The CIA World Data Banks are widely distributed and used for many mapping purposes. They consist of decimal latitude and longitude pairs. World Data Bank I at a source map scale of 1:12,000,000 consisted of 100,000 x, y coordinates. World Data Bank II was digitized at 1:3,000,000 and has about six million coordinates. Neither database has a topological structure, although topologically structured versions exist in several proprietary formats. World Data Bank II has several topological inconsistencies. Each major coastline simply continues as a single line until it meets a significant point, such as a major national boundary, or closes on itself. The same is true of the political and hydrological data. The beginning of a new line is the only break from the sequence of binary-encoded pairs of latitude and longitude in degree-minute-second format. This structure, or lack of it, was the origin of the term "cartographic spaghetti." During this project—at a conference dinner table—the data were described as being "no more structured than spaghetti on a plate." Ever since, the entity-by-entity string structure has been alternatively known as "cartographic spaghetti."

An important new dataset, which largely replaces the World Data Banks, is the Digital Chart of the World (DCW). DCW is a digital version of the Defense Mapping Agency's Operational Navigation Chart (ONC) and Jet Navigation Charts (JNC) at a scale of 1:1,000,000. The data are digital vectors in string structure, formatted in a manner very similar to the Vector Product Format (VPF) of the NATO DIGEST standard for digital data exchange covered in Chapter 6. The data cover some 14 layers, including coastlines, transportation, hydrology, contours, vegetation, inhabited places, and place names. The data are stored and distributed on four CD-ROMs, which include software for viewing the data files on IBM-PC compatible microcomputers. Distribution of the data is via the U.S. Geological Survey's Earth Science Data Centers. Information is available by calling 1-800-USA-MAPS in the United States and Canada.

DCW has some known problems that are emerging as the data are used for more and more applications. These include (1) problems of discontinuities at tile boundaries, because the data are tiled by latitude and longitude cells; (2) rings containing loops, such as an outer ring connecting to an included hole; and (3) rings that cross themselves. The size of the files, and the data being split across four CD-ROMs, means that queries and even simple draw operations on a microcomputer are extremely slow. Several efforts are under way to convert DCW to formats other than VPF, such as Arc/Info and DXF, and software vendors now offer their own software for DCW display.

The significance of the DCW for cartography cannot be understated. This single data set is likely to form the base layer for mapping around the world and in many ways is a fulfillment of the international efforts aimed at producing standardized world maps at this scale. The scale is particularly useful for integrating the broader satellite coverage such as AVHRR and some EOS data, and the detailed mapping coverages of the various nations of the world.

5.4.3 Formats at the Defense Mapping Agency

The Defense Mapping Agency (DMA) is the primary producer of digital cartographic data products for support of the U.S. military. The digital data form part of the Digital Landmass System (DLMS) and are made up of two parts, the Digital Feature Analysis Data (DFAD) and the Digital Terrain Elevation Data (DTED). Together, these data sets consist of about 25,000 magnetic tapes at 1,600 BPI (bits per inch, a tape data density), structured sequentially. This data set will eventually consist of 10^{19} bits of data. Overall, the DLMS contains information about terrain, landscape, and culture to support aircraft radar simulation, map and chart production, and navigation.

The DTED data contain elevation data sets in latitude, longitude, and elevation format that are used to generate elevations at 3-arc second intervals. These data are referenced to mean sea level as integer elevations rounded to the nearest meter. These data have been produced from contour maps and from stereo air photos, and have been enhanced with streambed and ridge information to preserve the topographic integrity.

The DFAD data contain descriptions of the land surface in terms of cultural and other features (forests, lakes, and such) stored as point, line, and area data. These data are lists of, or single latitude/longitude pairs and accompanying tables of attribute information. For example, a water tower would be stored in DFAD as a point and the attributes would include feature height and structure type, while a bridge may be recorded as a line segment. Linkage between the attributes and the cartographic data is maintained by header files.

The land features are coded at two levels. Level I data, with a resolution of about 152 meters (500 feet), contain large-area cultural information, such as surface material category, feature type, predominant height, structure density, and percentage roof and tree cover. Cultural features are digitized in a planimetric linear format, often called standard linear format (SLF). Standard linear format has been used as an internal digital cartographic data exchange format for the DMA.

Level II data are far more detailed, showing small area cultural information, and match the DTED data at 1-arc second resolution, about 30 meters (100 feet). These data are produced by manually digitizing large numbers of maps, charts, and air photos. The Level I data-bases will cover about 82 million square kilometers when complete. Level II data are digitized only for a limited number of areas of interest.

5.4.4 Formats at the National Ocean Service

Under the National Ocean Service (NOS) falls the Office of Coast and Geodetic Survey, and the National Oceanic and Atmospheric Administration (NOAA). These organizations produce two major types of cartographic products: nautical charts depicting hydrography and bathymetry and aeronautical charts which consist of visual flight charts and instrument flight charts. This division of the U.S. federal government conducts large-scale production of digital cartographic data, and some significant data sets are available.

The Nautical Charting Division of NOS offers digital shoreline data, and some aids to navigation. These data sets cover the shorelines of the coterminous United States, Puerto Rico, the U.S. Virgin Islands, and Hawaii, all to nautical chart accuracy standards. The data are available at small or large scale. The large-scale data have been digitized from the largest scale NOS nautical charts and plots of each part of the coastline, and they show by vectors the mean high-water mark. The shoreline vectors are continuous lines, but are broken where rivers flow into the ocean and where the shoreline cannot be precisely located. Each data set consists of between 100,000 and 300,000 points, and the data sets have been edge matched between data sets. Nineteen data sets cover the areas mentioned above, with approximately 20 more planned for Alaska. Three small-scale sets—East Coast, Gulf Coast, and West Coast—will be supplemented with an additional seven for the Great Lakes and Alaska.

The Nautical Charting Division of NOS has also embarked upon an ambitious program to convert all nautical charts to digital form as part of its ECDIS (Electronic Chart Display and Information System). As was the case at the USGS, these digital files were initially to be used to update and support standard printed map products. The newly evolved purpose is for these digital charts to be used by automated navigation systems for the piloting of ships and aircraft, as they now are. Efforts are under way to ensure that other nations of the world share data in a common format, thereby ensuring global standardized nautical charts for all the world's oceans.

The small-scale data are digitized from charts at 1:250,000 and show more generalized shorelines as a result. Records are 80 ASCII characters per record, written sequentially, with each point using one record. Each coordinate point is recorded as degrees, minutes, and seconds (to the nearest thousandth of a second) of latitude and longitude, with full geodetic datum, scale, source, date, and feature code reference. In addition, separate files contain point data aids to navigation, such as the locations of wrecks, fixed and floating navigation aids, and obstructions. These data are stored in a similar format to the shoreline data and contain about 30,000 points per data set. Between the two data sets, all the data for the nautical charts are geocoded and distributed.

One final and notable digital data set is the global elevation data available from the National Geophysical Data Center, part of NOAA. These data are surface elevations and ocean depths, at 5 minutes of increment for latitude and longitude, digitized from navigational and aeronautical charts by the U.S. Navy and known as ETOPO5. The entire data set contains 2.3 million observations and is distributed both as the whole world and as segments by continent. The distribution medium is nine-track magnetic tape or floppy disk. In the recent past, the National Geophysical Data Center has released numerous digital map data sets on CD-ROM, most recently including detailed bathymetry of the ocean and land-surface topography as well as geodetic and magnetic data for the earth's surface.

5.4.5 Digital Image Formats

Digital imagery is increasingly an important source of data for integration with digital vector maps in GIS. Two major types of digital imagery are of cartographic importance: scanned air photos and remotely sensed satellite and aircraft data. Numerous government agencies provide image data, including the National Aeronautics and Space Administration (NASA), the U.S. Geological Survey, and the National Ocean Service. Similarly, private companies distribute data from commercial land remote sensing programs, including EOSAT and SPOT.

NASA owns and operates numerous instruments and satellites for the collection of Earth data that are suitable for mapping. A very large amount of data from aircraft and spacecraft for Earth and the planets and their moons are available for cartographic purposes. The bulk of the data is in the public domain, and is available at the cost of reproduction to users. NASA maintains a central directory system for archiving and searching its data, which are scattered around the major NASA centers around the United States. The NASA Master Directory is covered in Chapter 6, because it is network accessible.

As part of the Mission to Planet Earth, NASA is building the Earth Observation System (EOS) (see Price et al., 1994). EOS data, from a very large array of instruments including those suitable for land-surface mapping, will be distributed through the NASA EOSDIS (EOS Data and Information System). The system will consist of a number of Distributed Active Archive Centers (DAACs), located mostly at NASA centers. The Land Surface Processes DAAC is under construction at the EROS Data Center in Sioux Falls, South Dakota. DAACs will be publicly accessible through the Internet and will support browsing and file transfer to geographic information scientists. An immense reserve of remotely sensed data will soon be available for mapping from these centers.

The USGS operates the central archive in the United States for the Landsat and AVHRR (Advanced Very High Resolution Radiometer) data provided by U.S. government mapping satellites.The data is stored on magnetic reel and cartridge tape at the USGS National Mapping Division's EROS Data Center in Sioux Falls, South Dakota.

All public domain Landsat data, and a large variety of other image format data, are available from this site. The EROS Data Center serves as the distribution point for many other types of digital data, such as DEMs. AVHRR data are available on an hourly basis, with 4- and 1-kilometer ground resolution. As such, they are used for intermediate and small-scale mapping of global, continental, and regional areas, including the United States. Sample data and useful overlays such as topography are available on CD-ROM.

The USGS also provides two other sets of digital image data of interest to cartography. These are the SLAR (Side Looking Airborne Radar) images, corresponding to 1:250,000 coverage of the United States, as part of an ongoing radar mapping program, and the new Digital Orthophotographic Quadrangle (DOQs).

The DOQ is likely to become a primary source of map information in the United States, especially for map updating and when a high-resolution map reference base is necessary. Distribution is by CD-ROM in quadrangle format, with four 1:12,000-equivalent quarter quads making up areas consistent with the USGS 7.5-minute series. The files are JPEG compressed gray scale (single band) images of about 50 megabytes apiece, with a ground equivalent resolution of 1 meter. The USGS, with assistance from the U.S. Department of Agriculture, plans to map the whole country in the DOQ format and to update the coverage on a five-year cycle.

Finally, an important source of up-to-date information on weather and atmospheric conditions is the GOES weather satellite data. Although this data are of very poor spatial resolution and is grossly distorted by Earth's curvature, its ready availability has made it popular with navigators and weather forecasters, including television stations. The data are network accessible; access is discussed in Chapter 6.

5.4.6 Industry "Standard" Formats

Several industry standards have emerged to satisfy the immediate data transfer needs in computer cartography. Although none of these data formats is accepted by national or international organizations as transfer standards, proprietary standards are nevertheless extremely common in the work environment.

At the initial level, many software packages maintain their own format and file structures for data in use by the software. Arc/Info, for example, uses protected file formats and structures the files according to projects, geometry, and relationships between the data. All files containing data for the INFO relational database manager, for example, must be in a separate directory under the project directory. Similarly, most commercial systems use proprietary formats to optimize storage efficiency, increase storage, or optimize a system around a particular piece of hardware. Most packages that use proprietary formats contain modules that read (and sometimes write) many of the formats covered in this chapter. Without this capability, users would be frustrated because they would be unable to bring in new cartographic data.

Some proprietary formats have emerged as common to many systems and therefore serve as *de facto* exchange standards (Kay and Levine, 1992). None of these is optimized for cartographic applications, such as allowing the exchange of cartographic geometry, topology, and attributes. Nevertheless, many systems allow exchange of attributes from statistical packages such as SAS, from spreadsheets, or from database packages, thus allowing the reassembling of the information after the transfer. Many systems do not handle the transfer of topology at all, requiring its reconstruction when the map is transported between computers.

For vector data, the three most common industry standards are AutoCad DXF, PostScript, and HPGL. DXF, for AutoDesk's Digital eXchange Format, is broadly used in the computer-assisted drafting and design field. The format consists of a large number of default ASCII text labels, accompanied on the following line by numerical values for these fields. At the core is a list of x and y values defining the drawing, in this case forming the lines on a map. DXF supports structuring by layers, a concept familiar in GIS, but in a graphic rather than a topological sense. Both PostScript and HPGL are really plotter formats, designed to allow hardware devices (laser-jet printers using the Adobe software and fonts, and Hewlett Packard printers and plotters) to translate x and y increments into firmware plot commands. Typically, these formats involve a base level with ASCII codes and coordinates corresponding to move and draw commands and other structured commands which establish the plotter size, pen colors, text font, and size and so forth.

For raster data, a particularly rich set of industry standards has developed. Among the leading formats are Targa, PICT, TIFF, GIF, PCX, JPEG, and Encapsulated PostScript. Many scanners, which generate a raster format, and bitmap editors such as PC Paintbrush, SuperPaint, Adobe Photoshop, and IslandPaint, import and export combinations of these formats. A few packages can import and export both raster and vector formats, such as CorelDraw! and Adobe Illustrator. In addition, several packages are available to translate between these formats, such as Freedom of the Press and Image Alchemy.

Each of these formats is similar, with the only exceptions being those using compression schemes (in this case GIF and JPEG). Many of the formats work across bit depths, so that a 24-bit image will be displayable in binary, for example. Targa, PCX, and TIFF files are similarly structured, in that they consist of a single file containing a file header, a color map, and a set of color indexes. The color map is a set of Red, Green, and Blue (RGB) intensities onto which the data values are mapped. Thus color index 10, for example, may consist of the color pure red (red = 255, green = 0, blue = 0). As the image grid is extracted from the file, all values of 10 are then given these RGB values for display. In systems that advertise a selection of 256 possible colors from a "palette" of over 16 million, the 16 million comes from being able to index $256 \times 256 \times 256$ colors, but only 256 at a time since the data array itself is only eight bits deep.

JPEG and GIF formats implement compression strategies, of which the most sophisticated is that used by JPEG. JPEG is able to compress files with some degree of redundancy in their underlying data many times, for example 50:1. JPEG also allows partial recompression,that is, lossy compression, in which some of the image data are lost to further compress the data. This is a significant problem for many cartographic applications, where loss-less compression is preferred. The lineage of these file structures varies. Targa files were designed to support a particular graphics display board for microcomputers. GIF files were designed for shipping images over networks, in particular the CompuServe system.

PCX is proprietary to Zsoft, but is supported (along with the BMP Windows bitmap format) in the Paintbrush drawing accessory program which comes free with Microsoft's

Windows operating system for IBM-PC compatible microcomputers. TIFF formats have been more popular for use on Apple Macintosh computers, in software packages such as Adobe Illustrator, and for scanners.

For specific needs, these formats are often more than adequate. Most major software packages for computer cartography will deal with one or more of these formats. Rarely are these internal or programming formats, however, but instead they are invoked when data are to be stored as external binary or ASCII files on disk. Because none of the formats is a national or international standard, their future is not guaranteed. Archiving data for long periods in these formats is likely to entail a significant risk.

To counteract the problems associated with long-term storage and exchange of data over noncompatible systems, several formal standards efforts have been initiated and in a couple of cases completed. These standard formats represent the best possibility for digital cartographic data to be both broadly distributed and exchanged between cartographers and geographic information scientists. The standards efforts are covered in detail in the Chapter 6, and one standard, the Spatial Data Transfer Standard, is used in the development of a set of C language data formats, models, and structures for use in computer programming.

5.5 REFERENCES

Defense Mapping Agency (1985a). *Standard Linear Format.* Washington, DC: U.S. Government Printing Office.

Defense Mapping Agency (1985b). *Feature Attribute Coding Standard.* Washington, D.C: U.S. Government Printing Office.

Department of the Interior, U.S. Geological Survey (1985a). *Digital Line Graphs from 1:100,000-Scale Maps, Data Users Guide.* National Mapping Program Technical Instructions, Data Users Guide 2, Reston, VA.

Department of the Interior, U.S. Geological Survey (1985b). *Geographic Names Information System, Data Users Guide.* National Mapping Program Technical Instructions, Data Users Guide 6, Reston, VA.

Department of the Interior, U.S. Geological Survey (1986a). *Land-use and Land-cover Digital Data from 1:250,000- and 100,000-Scale Maps, Data Users Guide.* National Mapping Program Technical Instructions, Data Users Guide 4, Reston, VA.

Department of the Interior, U.S. Geological Survey (1986b). *Digital Line Graphs from 1:24,000-Scale Maps, Data Users Guide.* National Mapping Program Technical Instructions, Data Users Guide 1, Reston, VA.

Department of the Interior, U.S. Geological Survey (1987a). *Digital Line Graphs from 1:2,000,000-Scale Maps, Data Users Guide.* National Mapping Program Technical Instructions, Data Users Guide 3, Reston, VA.

Department of the Interior, U.S. Geological Survey (1987b). *Digital Elevation Models, Data Users Guide.* National Mapping Program Technical Instructions, Data Users Guide 5, Reston, VA.

Guptill, S. C., ed. (1990). *An Enhanced Digital Line Graph Design.* Department of the Interior, U.S. Geological Survey Circular 1048. Washington, DC: U.S. Government Printing Office.

Held, G. (1983). *Data Compression: Techniques and Applications, Hardware and Software Considerations.* New York: Wiley.

Kay, D. C., and J. R. Levine (1992). *Graphics File Formats.* Blue Ridge Summit, PA: Windcrest/McGraw-Hill.

Price, P. D., et al. (1994). "Earth Science Data for All: EOS and the EOS Data System." *Photogrammetric Engineering and Remote Sensing*, vol. 60, no. 3, pp. 277–285.

6

Access to Spatial Data

6.1 FINDING SPATIAL DATA

When the users of cartographic data are municipalities, utilities, or state and local governments, there is often a significant amount of geocoding to be done before a mapping project can begin. This data capture can be extremely costly, however, and increasingly these mapping organizations are looking toward the use of digital data that already exist, even if that data acts only as a base upon which to expand. If the cartographer is seeking data for a map that are likely to already exist in digital form, then the problem is not to geocode, but to find and access the data.

As we have seen in Chapter 5, there is no shortage of existing data suitable for analytical and computer cartography. This situation means that new data need to be geocoded only in those circumstances where new layers or themes are needed, where new data have to be integrated with existing maps, or when digitally unmapped areas are to be used. Rather than actually dealing with the specifics of data storage formats, data models, and data structures, the computer cartographer is often faced with simply finding *where* data sets reside, whether they are *accessible,* and through which storage media distribution is possible.

As more and more data are demanded by computer mapping and information management, the more the advantages of digital data become apparent. Digital data can flow easily across computer networks, making networking the second computer revolution in cartography and making the problem of distribution and storage vastly simpler. This chapter discusses some of the means to find and access spatial data, a task that is becoming more and more common. We examine the role of map libraries, the role of the national spatial data infrastructure, and the power of the Internet, and we conclude with a discussion of spatial data transfer standards.

6.1.1 Using a Map Library

The search for paper maps is often conducted in a library. Libraries most likely to carry maps and support cartographic research are the research libraries in the largest cities and those attached to major universities. Map librarians make use of computer networks to share information and conduct searches, and they are increasingly making census and other digital maps available both in libraries and via networks.

Map librarians have faced difficulties in preparing for the digital cartographic revolution due to a text-only library tradition, tight budgets, and an inability to educate sufficient numbers of map librarians in digital mapping techniques. Nevertheless, lower costs for hardware and software, new technology such as CD-ROM, extensive demand for Census Bureau TIGER data, and a new awareness of the need for digital map libraries have led to significant changes. Some pioneering libraries began the transition in the mid-1980s, while now about half of the nation's libraries either retrieve and store digital map information directly or have used network links to establish a new working relationship between map libraries and their users. This transition has been helped in some cases by educational programs, such as the American Research Libraries initiative to educate and equip libraries in the provision of spatially referenced data in all formats. This program has led over 70 libraries into the digital era, providing training, hardware, and software.

The map librarian now plays a role as a broker of spatial information, linking the right data with the user and perhaps even providing the first software-based display of the data and a hard-copy computer-generated map. The network link has also changed the view of the map librarian, from that of an acquirer of a copy of a map for a user to that of a custodian of a section of unique data on a network for access by users of all libraries. Network access to library catalog information has stressed this distributed database model.

In addition, commercial companies often sell cartographic data and will conduct searches. Landsat imagery, for example, from EOSAT, can be searched for and browsed using an on-line database. Major U.S. government programs such as the Global Land Information System and NASA's Earth Observation Systems have evolved similar concepts, including image browsing by network.

6.1.2 The National Spatial Data Infrastructure

The U. S. Geological Survey makes its data available through the Earth Science Information Centers, where questions about digital map data availability within the federal government are answered. Some states, such as South Carolina and Wisconsin, have data clearinghouses that make data available to users. There is also a large variety of journals, newsletters, and information sources to suggest data sources.

A prevailing attitude in the United States has been the concept of digital map data, at least raw "base" map information, as a public good. The growth of geographic information science, with its powerful suite of tools in geographic information systems technology, has taken place largely because of the abundance and low cost of data. This is particularly obvious when the system in the United States is compared with full or partial cost-recovery systems, such as that used in the United Kingdom by the Ordinance Survey (Rhind, 1992). In many applications of computer mapping and GIS, data production or conversion often accounts for over 80% of the cost. With free or low-cost data, the cost of projects obviously drops significantly.

At the national level in the United States, an initiative is now under way to develop the NSDI, or National Spatial Data Infrastructure (National Research Council, 1993). This system would make digital spatial data broadly available to the public, probably over computer networks such as the Internet or through "published" CD-ROMs. In addition, the system would coordinate the mapping activities of government and other mapping agencies, to reduce redundancy and to make map data sets match each other across scales and at their boundaries.

The ability to search and retrieve data quickly will give rise to new cartographic applications, such as the mapping of natural disasters as they occur, the use of computer cartography in search-and-rescue, the use of in-vehicle navigation systems, and the development of many other possible uses, especially when coupled with the position-locating capabilities of the Global Positioning System. Both analytical and computer cartography have a considerable amount to gain from the National Spatial Data Infrastructure, as it has been proposed.

Early systems that had some of these characteristics are NASA's Land Pilot Data System, USGS's Global Land Information System, and NASA's EOSDIS Version Zero. What has changed significantly since the origins of these systems is the emergence of the Internet as a highly effective mechanism for both data searching and data distribution.

6.2 USING THE INTERNET AND FTP

There is little doubt that the vehicle of choice for the distribution and searching of spatial data is the Internet, a network of computer networks available worldwide (Krol, 1993). The Internet is a network of computer networks and is accessible to all users through a computer that is attached to the system. In addition to e-mail (electronic mail) worldwide, the Internet allows the users of computer mapping systems access to the existing National Spatial Data Infrastructure, which is assembling itself spontaneously at a remarkable rate of speed. The Internet contains several information and data sources of great value for cartographers, primarily network conference groups, file transfer mechanisms, and network search capabilities.

6.2.1 Network Conference Groups

In terms of information, several network conference groups exist for cartographers, and others focus on specific software products. In each case, a list is subscribed to by sending a mail item to the Internet address in the following way (which uses Unix Mail):

```
mail LISTSERV@address
Subject:
subscribe LIST-L Your Name
<control>-D
```

where the `address` is the Internet address, the e-mail system query for a subject is left blank, the list name and your name follow the subscribe command, and the message consists only of these items. Alternatively, the news reader system can be used, which "screens" your mail and allows it to be read selectively. Other systems work similarly.

Most lists have an archive of FAQs (frequently asked questions), which should be located and read *before* you begin using the list. It is a good idea to monitor a list as a passive reader for a few days before becoming active. A common mistake for new users is to send subscription requests to the list name instead of the list server. This error is extremely frustrating to newsgroup users, who are then bombarded with every single start and stop request for subscriptions. **This error is to be avoided at all costs. It is a severe breach of network etiquette.**

MAPS-L at `uga.cc.uga.edu` is a medium-volume moderated newsgroup for map librarians and often has interesting locational queries. A Canadian equivalent is CARTA, at `sask.usask.ca`. GIS-L at `ubvm.cc.buffalo.edu` is a high-volume newsgroup for GIS users. It is often highly technical, but is frequently lively and contains much of interest to analytical and computer cartography.

This list is cross listed at `comp.infosystems.gis` on Usenet, where e-mail tools make monitoring messages and screening by subject easier. As do many lists, this list has an `FAQ` (frequently asked questions) `list`. To get the FAQ list one can anonymously ftp (see Section 6.2.3) from `abraxas.adelphi.edu` (file `/pub/gis/FAQ`) or send a message to `gis-faq-request@abraxas.adelphi.edu`.

Other GIS lists include TGIS (temporal issues in GIS), UIGIS (user interface issues), and CPGIS (Chinese Professionals in GIS). INGRAFX is a very low volume list at `psu-vm.psu.edu`. It is a list designed for those interested in issues and research in mapping and visualization. IMAGRS-L at `csearn.bitnet` is a list devoted to remote sensing and has some spillover into cartography.

Support lists for specific software packages can be used for help in implementing software. These include `grassu-request@zorro.cecer.army.mil` for the U.S. Army Corps of Engineers GRASS GIS and image processing package (both software and data are available), and `idrisi-l@toe.towson.edu` (send subscribe message to `MAILSERV`, not `LISTSERV`. The NSDI discussion list is `NSDI-L`, with the subscription message to `listproc@grouse.umesve.maine.edu`. Some commercial companies also use lists for technical support.

Finally, a note of caution. Network lists are somewhat ephemeral. Lists evolve, move, and sometimes die. The use of a network browsing tool such as mosaic to locate active lists is recommended.

6.2.2 Network Searching

It is also possible to conduct searches over the network. Searches are often conducted using network browsing tools such as ARCHIE, GOPHER and WAIS. An overview tool, `mosaic` (Ritter, 1994), allows access to each of the other searching systems, acts as a gateway into systems, and allows menu and window-based queries across the network (Figure 6.1). These tools allow a user to search publicly available files located on servers.

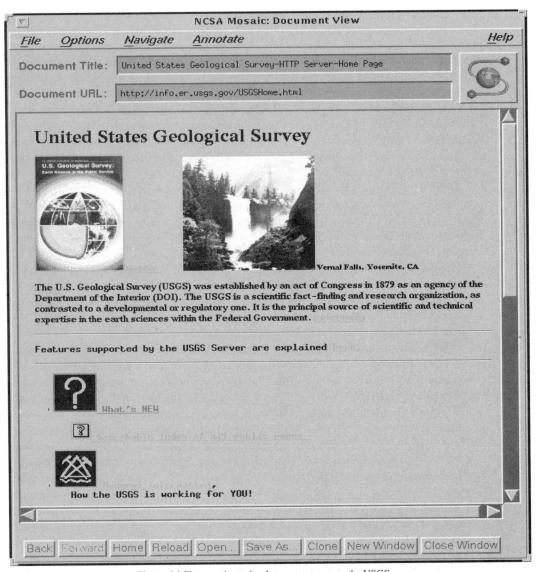

Figure 6.1 The mosaic gopher Internet gateway to the USGS.

Servers are file archives located on the network that are accessible over the network either by remote login or by the file transfer protocol FTP. The user can also search directly, if a specific Internet address is known.

NASA maintains a Master Directory of all its data sets that is searchable by latitude and longitude. It can be used by first setting your workstation or terminal in vt100 emulation mode (in Unix, `setenv TERM vt100`). The system is comprehensive, and can be used to access several other systems and many individual sites. To use the system, `telnet` to `nssdca.gsfc.nasa.gov` (`128.183.36.22`). At the prompt, enter

NSSDC, for `username`, enter NSSDC, and then select option 1 from the top-level menu.

The Master Directory allows browsing and queries to many other databases, including those of the EROS Data Center and the Pilot Land Data System. It is likely that this mechanism will also contain EOSDIS, NASA's Earth Observing System Data and Information System, which will be used to archive and distribute data from many satellites suitable for mapping, including LANDSAT 7. Another search directory of use is the USGS GLIS database (Global Land Information System), which includes index references to many USGS map products including DLGs and DEMs and satellite imagery such as AVHRR (`glis.cr.usgs.gov` or `152.61.192.54`).

6.2.3 Network File Servers

In many cases, digital map data are available from individual file servers. The means by which access is made is to use the *File Transfer Protocol* (FTP). Using this protocol allows one to `login` to a guest account on the file server, usually using the login name of `anonymous`. By courtesy, it is common to use your own Internet address as the password, thus leaving a trail for the owner to trace who is accessing the databases. The individual servers can be located by using the network search tools, or they are announced in network news groups.

Many of the cartographic data sets in the formats described in the previous chapter can be found by searching and then transferring via FTP. Table 6.1 contains a set of such internet addresses for servers. From the USGS server, for example, the 3-arc second DEMs, the GIRAS, the ETOPO5 , and several other data sets are available. The availability of data is continuously increasing. News about new data listings and postings is usually available at a top level menu. Many of the data sets are compressed to make them easier for network transfer, usually with the UNIX compress facility, which results in a file with a "`.Z`" filename suffix. These files should be FTP transferred using the binary format, rather than as ASCII. The FTP command `binary` selects binary file transfer.

It is likely that the Internet will remain a primary means of data distribution and will expand considerably in the future. A significant amount of data and in many cases software as shareware can be downloaded for microcomputers and workstations from the file servers attached to the network. As a network user, you have an obligation to return useful or interesting software or data of use to other users via the network.

6.3 CARTOGRAPHIC DATA TRANSFER STANDARDS

Moving map or any other data over a network requires that the recipient be able to use the data when it arrives. In Chapter 5, we saw that two types of data formats have developed for cartographic data: those that are the result of data production by an agency with a need to distribute the information, often along the lines of an existing paper map product, and the industry standard formats, which have evolved out of casual transfer of image and other data. Broad-scale transfer of data over networks is time-consuming and can rather quickly fill a disk. Proliferation of formats for exchange have resulted in costly and duplicative efforts to write file format converters and translators.

Table 6.1 Network File Servers with Cartographic Data

Address	Login and Directory	Contents	Comment
`sseop.jsc.nasa.gov`	`login anonymous`	NASA space shuttle images	
`ames.arc.nasa.gov`	`login anonymous cd pub/GIF`	NASA space shuttle images	
`vmd.cso.uiuc.edu`	`login anony- mous, cd wx`	GOES weather images	Available every hour
`inidata.ucar.edu`	`login anonymous cd images`	GOES weather images	Available every hour
`aurelie.soest.hawaii.edu`	`login anonymous, cd /pub/avhrr/ images`	GOES weather images	Available every hour
`hanauma.stanford.edu`	`login anony- mous, cd pub/ World_Map`	The CIA World Data Bank I and II	Includes soft-ware for draw-ing the files, start with the README file
`martini.eecs.umich.edu`	`telnet`	Latitude, longi-tude and elevation for U.S. cities	
`glis.cr.usgs.gov`	`telnet`	Master directory and descriptions of most types of data	
`alum.wr.usgs.gov`	`anonymous, cd pub/maps`	Geology maps	Includes dis-play software

Although many different organizations, both private and public, have developed their own formats for digital cartographic data, the disparities between formats have become disadvantageous only since the ability to share and distribute these data has become possible, especially across Internet but also on local area networks. Distributed computing has evolved to a level where databases normally reside at host locations and are made network accessible to the actual applications,that is, the mapping programs, which actually use them. Thus a map data set could remain on disk in one location, but be mapped and displayed at another.

Designing a system to accept data from two formats needs either a format translator or two translators into a third common or "standard" format. A system to convert data between 10 formats, however, must have either 10 times 9 translators or 10 forward and 10 reverse translators to and from a single, universal standard. In the long run, clearly a commonly accepted standard is most desirable. There have been some pioneering efforts to standardize digital cartographic data formats. Among the standards are several developed for data exchange within government agencies. These agencies include the Committee on Exchange of Digital Data (International Hydrographic Organization), the Federal Geographic Exchange Format of the Federal Interagency Coordinating Committee on Digital Cartography, the Geographic Data Interchange Language of the Jet Propulsion Laboratory (NASA), the Standard Digital Data Exchange Format of the National Ocean Service, plus those discussed previously (Langran and Buttenfield, 1987).

Several standards for digital map data have emerged as a result of these efforts. The Spatial Data Transfer Standard, approved as a Federal Information Processing Standard (FIPS) in 1992, will be considered in the following section. Parallel efforts have resulted in DIGEST from NATO and DX90 from the International Hydrographic Organization. Some other countries, such as Australia and Germany with ATKIS, have initiated standards efforts. In addition, DLG-E, mentioned in Chapter 5 as a revision of the USGS's Digital Line Graph system, can be considered a standard, although it clearly had an influence on SDTS (Guptill, 1991).

DIGEST (Digital Geographic Exchange Standard) grew out of the military's and the Defense Mapping Agency's standardization efforts, such as the Standard Linear Format (see Section 5.4.3). After several versions, the standard received Military Standard status in 1992. DIGEST has several versions and parts. DIGEST-A is feature based, while DIGEST-C uses a relational data model (see Chapter 9). DIGEST-C is also known both as VRF (Vector Relational Format) and as the Vector Product Format (VPF). The Digital Chart of the World is stored in VPF on CD-ROM. DIGEST-A uses a telecommunications standard (ISO 8824) in the place of the static medium file storage standard ISO8211, which is also used by SDTS.

DX90 is a digital exchange format sponsored by the Committee for the Exchange of Digital Data and it is accepted by the International Hydrographic Organization. It is intended to support the use of digital map data for charting, safety, and navigation at sea. DX90 is the popular name for IHO special publication number 57, the Transfer Standard for Spatial Data. This standard is under revision and is undergoing a learning process as the various nations of the IHO begin to use the standard to exchange data.

Out of these standardization efforts grew the National Committee on Digital Cartographic Data Standards, more commonly called the Moellering Committee, which was formed in 1982 under the auspices of the American Congress on Surveying and Mapping as a result of a request from the USGS. In 1987 this committee completed the first draft for a new set of common data standards for digital cartographic data. These standards have now been approved by the National Institute for Standards and Technology as a FIPS.

The standards have substantially influenced computer cartography. This has been due to their 10-year evolution, and the substantial involvement of the user communities in government, academia, industry and other interested parties. The Spatial Data Transfer Standard, as the final standard is known, provides a uniform and consistent terminology, a set of definitions, and a set of formats for information exchange that now form a mandatory framework for federal data transfers.

The scope of the standard, however, goes well beyond federal needs, and the standard has already influenced software producers, state and local governments, and the activities of all cartographers. As such, the SDTS forms the remainder of the discussion for this chapter, concluding with a set of C language programming data structures that are consistent with the terminology of the standard.

6.4 THE SPATIAL DATA TRANSFER STANDARD

The Spatial Data Transfer Standard, issued August 28,, 1992, as FIPS 173 (Department of Commerce, 1992) culminated a 10- year standardization effort that began as an attempt to standardize digital cartographic data and ended with a broad standard with wide scope for application to all spatial data and spatial data transfers. The purpose of the standard is to promote and facilitate the transfer of digital spatial data between dissimilar computer systems. The standard provides a common mechanism for the transfer of data, provides a set of clearly defined spatial objects and relationships to represent real-world spatial entities, and provides a transfer model to facilitate the translation of user-defined objects into the standard.

The standard consist of two parts. Part 1 contains the logical specifications, including concepts, object definitions, data quality, and transfer module specifications. Part 2 consists of a dictionary of entities, that is, definitions of the cartographic features to be encoded in the standard. The spatial data transfer component of the standards attempts to meet the requirements for moving digital cartographic data between systems. The USGS serves as the maintenance authority for this standard, which lists a series of steps to be taken to attain conformance.

The definitions and references within SDTS are a systematic attempt to define a set of cartographic primitives as zero-, one-, and two-dimensional objects, with which digital cartographic feature representations, that is, symbols and maps, can be built. It should be noted that the standard definitions are restricted to these object definitions and do not propose any particular method or type of symbolization. The language of the geographic entity and object is used—concepts that are used throughout this book.

The section on data quality proposes that digital cartographic data include a quality report, either as a paper document or as part of the data set. Elements of the report include lineage of the data (source and modification history, for example), positional accuracy, attribute accuracy, logical consistency, and completeness of the data. These requirements seem critical to the long-term survival of data sets, especially within GISs, and also are of critical importance for the provision of an assessment of reliability for a particular map product made from digital cartographic data.

The final section of the standard relates to cartographic features. Central to this section is a complete set of cartographic entity and attribute definitions. This set of features in intended to be as complete as possible, though provisions for update are made. This means, for example, that when a digital data set references a "water tower," there is complete agreement on what constitutes a water tower. For example, the cartographic entity CHURCH is an included term number for the entity BUILDING, defined as "a permanent walled and roofed construction."

Two other important sections of the proposed standard document are of direct interest. First, under the cartographic objects section, a glossary is included. Second, , the standard also includes a bibliography, to which the reader is referred for additional information.

Implementation of the SDTS has proceeded using the mechanism of the profile. A profile is a subset of the standard, reflecting the demands of a particular type of geographic data. The first profile, now added as part of the FIPS standard, is the Topological Vector Profile (TVP). Specification of the TVP has allowed many U. S. federal data sets to be restructured into SDTS, including SDTS-DLG and SDTS-TIGER. The maintenance authority for the standard, the USGS, has released public domain software designed to read and write data into the SDTS-TVP file structures. Many software vendors are now incorporating SDTS-TVP readers into their programs. Information about the profiles and the standard in general are available over the Internet from `sdts@usgs.gov` and can be found using Mosaic, WAIS, or Gopher.

The second stage for SDTS is to implement the Raster Profile of SDTS. The raster profile will result in an extension to the FIPS standard in the near future. Therefore, the implications for the distribution of government image data, such as GOES and AVHRR data, as well as Landsat and future EOSDIS data, are significant.

6.5 REFERENCES

Department of Commerce (1992). *Spatial Data Transfer Standard.* FIPS PUB 173, Federal Information Processing Standards Publication, Computer Systems Laboratory, National Institute of Standards and Technology, Gaithersburg, MD.

Guptill, S. C. (1991). "Spatial Data Exchange and Standardization." Chapter 34 in D. J. Maguire, M. F. Goodchild, and D. W. Rhind, *Geographical Information Systems: Principles and Applications, Volume 1: Principles.* London and New York: Longman Scientific and Technical /Wiley.

Kottman, C. A. (1992). *Some Questions and Answers about Digital Geographic Information Exchange Standards.* Reston, VA: Intergraph Corp.

Krol, E. (1992). *The Whole Internet: Users Guide and Catalog.* Sebastapol, CA: O'Reilly and Associates,

Langren, G. and B. P. Buttenfield (1987). "Formatting Geographic Data to Enhance Ma-
nipulability." *Proceedings, AUTOCARTO 8,* Eighth International Symposium on
Computer-Assisted Cartography, Baltimore, March 29–April 3, pp. 201–210.

National Committee for Digital Cartographic Data Standards (1988). "The Proposed
Standard for Digital Cartographic Data." *American Cartographer,* vol. 15, no. 1, pp.
9–142.

National Research Council (1993). *Toward a Coordinated Spatial Data Infrastructure.*
NRC Mapping Sciences Committee, Commission of Geosciences, Environment and
Resources. Washington, DC: National Academy Press.

Rhind, D. (1992). "Data Access, Charging and Copyright and Their Implications for
Geographical Information Systems." *International Journal of Geographical Information Systems,* vol. 6, no. 1, pp. 13–30.

Ritter, M. E. (1994). "Mosaic and the World Wide Web." *Cartographic Perspectives,*
no. 19, Fall 1994, pp. 44–47.

Tosta, N. (1993). "The Data Wars: Part I." *Geo Info Systems,* January, pp. 25–27.

7

Spatial Data Structures for Computer Cartography

7.1 C LANGUAGE STRUCTURES FOR CARTOGRAPHIC DATA

The Spatial Data Transfer Standard has provided cartographers with a consistent set of terminology and concepts, known as a *data model*, around which data structures can be developed. Although most data structures are expressed in published form as logical models of data organization, in computer cartographic software they find their actual implementation as computer programming data storage mechanisms, what we will call here *program data structures*. These program data structures must relate to the goals of geocoding discussed in Chapter 4,; they must relate physically to the actual methods of storage and representation of digital data discussed in Chapter 5, and they also ideally should be simple and effective to use in any programming environment.

Many of the nonstructured computer programming languages did not support sophisticated data structures. More recent languages support these structures, especially the languages most likely to be used in most contemporary computing environments: Pascal and C. In Pascal, the "record" is similar to the "structure" in C, and the advanced student will be able to translate most of the computer programs discussed here between the languages with relative ease.

In the C programming language, the "structure" is declared in the following way. We take advantage of the C construct `typedef`, which allows us to name a structure as a type of storage object or data structure (Function 7.1).

Function 7.1

```
/* Example of a C language structure declaration
/* kcc 5-93 */
typedef struct {
    float first;
    int second;
} EXAMPLE;
```

To use this structure in a program, first a variable must be declared to have this particular structure as its type (Function 7.2).

Function 7.2

```
/* Declaration of a variable to be of type EXAMPLE
/* kcc 5-93
*/
EXAMPLE hold_data;
```

Each structure element is then used by linking subelements with periods. For example, to assign a value to the first element and print the value of the second, assuming they exist, the following program lines in Function 7.3 could be used.

Function 7.3

```
/* Example use of data stored in a structure (kcc 6-93) */
EXAMPLE hold_data;
hold_data.first = 3.1415927;
hold_data.second = 1;
printf("The first element of the structure is %f\n",
    hold_data.first);
printf("The second element of the structure is %d\n",
hold_data.second);
```

A structure can be multidimensional, as in the example Function 7.4.

Function 7.4

```
/* Example declaration of a multidimensional variable to be
of type structure kcc 6-93 */
typedef struct
    {
    float first; int second;
    } EXAMPLE;
EXAMPLE hold_data[5];
    for (i = 0; i < 5; i++)
        {
        printf("Structure element one sub %d = %f",i,
            hold_data[i].first);
        printf("Structure element two sub %d = %d",i,
            hold_data[i].second);
}
```

FORTRAN programmers should note that C language arrays all start at a count of zero The final concept for C language structures is that a structure can be part of the declaration of another structure. An example of this is shown in Function 7.5.

Function 7.5

```
/*  Example use of the POINT data structure  kcc 6-93
/* The first statement states that structure definitions
/* are in cart_obj.h in this directory */
#include "cart_obj.h"
/* Now declare a structure point, with 50 elements of type
     POINT */
POINT Point[50];
/* Print, for example, the 30th elements of point (indexed
 by 29) */
printf("(x,y) of 30th element is (%f,%f)\n",Point[29].x,
           Point[29].y);
```

The use of structures permits a high degree of organization in computer programs, especially because structures can be passed as arguments between functions or can be declared outside the function definitions to make them available anywhere within a program. To achieve this characteristic, somewhat similar to using COMMON blocks in FORTRAN, the structure definition can be placed together in a separate file. C supports such declarations, using the term *header file* for these files.

In the following declarations and functions, it is assumed that there exists a file called cart_obj.h that consists of declarations of the C language data structures to be used in this book. In each case, the structures are given uppercase names, so that when variables are declared to have this type of structure, they can use the same name in lowercase without confusion.

For example, the first declaration defines the zero-dimensional primitive cartographic object POINT. To declare a variable to be of type POINT, we can still call it Point (mixedcase, because GKS reserves some lowercase only words such as link) and perhaps give it multiple instances.

Thus a collection of points, a maximum of 50 for example, could be stored and used in a C language structure Point as shown in Function 7.5. Because the digital cartographic data standards divide the major cartographic object definitions into zero, one, and two dimensions, the same subdivision will be followed for the C language structures.

The following sections outline C language structures suitable for storing cartographic or spatial objects of zero, one and two dimensions. The purpose of these functions is to encourage the use in computer programs of data stored in a manner consistent with the SDTS data model. In each case, the objects are presented along with programming data structures which will hold data, and in some cases, example programs are given that use the structures to perform geographic operations.

7.2 ZERO-DIMENSIONAL CARTOGRAPHIC OBJECTS

The primitive object with one dimension is the *point*. SDTS specifies a point as a zero-dimensional object that specifies geometric location. The location is specified by a coordinate pair (x,y) or by a triplet (x,y,z). The rules for providing coordinates are discussed in the standards document in Appendix C "Spatial Address Encoding." A point is inherently a vector data element. Many program applications require the coordinates to be floating point numbers (for example latitude and longitude in decimal degrees). On the other hand, storing two or three floating point values for every location requires significantly more storage and computational power than using integers.

To accommodate both these formats all of the following functions use the C keyword typedef, which leaves the choice up to the programmer. It is important to remember in algorithms and programs which type is used. In the program examples in this book, neither type is assumed, and all programs perform type conversions where necessary. Careful attention should also be paid to whether negative values are appropriate, what the maximum values are likely to be, and whether there is a loss of precision involved.

Three special cases of a point are given, allowing the point to store an attribute associated with it, either an *entity*, a *label*, or an *area* identifier. The final zero-dimensional object is the *node*. The node is defined as a zero-dimensional object, that is, a topological junction of two or more links or chains, or an end-point of a link or chain. Examples are shown in Figure 7.1.

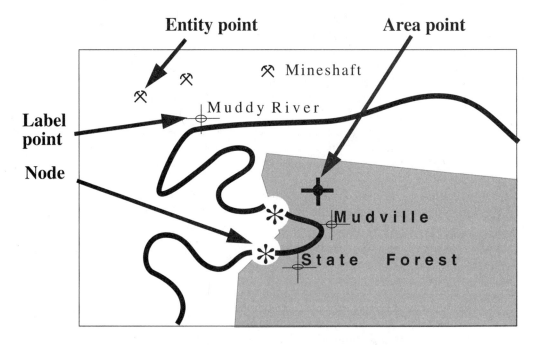

Figure 7.1 Zero-dimensional cartographic and geographic objects.

Function 7.6

```
/* -------------------------------------------
/* C language cartographic data structures
/* Based on FIPS 173 Terminology and data model
/* kcc 3-94 Revised from 5-88
/* -------------------------------------------
/* Define the constants for all objects */
#define MAXPTS 999
#define MAXROW   20
#define MAXCOL   20
#define MAXSTRING 400
#define MAXARC      100
#define MAXLINK     100
#define MAXCHAIN  100
#define MAXLABEL  20
#define MAXHOLE 4

/* Establish the coordinate type */
typedef coordinate {float | int }; /* NOT C, choose one */
/* Zero dimensional objects */
typedef struct {
      coordinate x;
      coordinate y; } POINT ;
typedef struct {
      char entity_type[MAXLABEL];
      char entity_attribute[MAXLABEL];
      char included_term[MAXLABEL];
      POINT Point;
      } ENTITY_POINT;
typedef struct {
      char label[MAXLABEL];
      POINT Point;
      } LABEL_POINT;
typedef struct {
      int polygon_id;
      POINT Point;
      } AREA_POINT;
typedef struct {
      int node_identifier;
      POINT Node;
      } NODE ;
```

The normal means by which structure definitions are made in C language programs is by placing them in a header file. The header files can then be included in each function that uses the structure, and the data are available to all functions in the program. A generic header file is built in this section to cover each of the objects defined in the Spatial Data Transfer Standards. This header file (Function 7.6 and continued as Functions 7.10 and 7.12) is called `cart_obj.h`, for the cartographic objects header file.

In the `cart_obj.h` header file, the initial set of statements (the `#defines`) allow the preprocessor to assign any constants to be used in the structure declarations. For example, instead of allowing a fixed value of 20 as the maximum number of points, the constant `MAXPTS` is used and is simply switched at compile time for the constant 20 by the C preprocessor. Because only those constants and structures which the program will use are loaded at run time, including the whole `cart_obj.h` file incurs no extra overhead.

An example of a point has already been given above. An entity point may contain a feature type of "church", identified in the feature code section of the data standards as entity type `BUILDING` (a permanent walled and roofed construction), included term `Church` (which has the attribute of `EXISTING`), for example at 1 degree 30 minutes East, 55 degrees 20 minutes and 45 seconds North.

Function 7.7a

```
/* Program segment to demonstrate an entity point
/* structure kcc 6-93 Revised from 12-88 funct 7.7a */
#include "cart_obj.h"
(void) show_entity_point()
{
    ENTITY_POINT Feature;
    Feature.entity_type = "BUILDING";
    Feature.entity_attribute = "EXISTING";
    Feature.included_term = "Church";

    Feature.Point.x = (coordinate) 1.3000;
    Feature.Point.y = (coordinate) 55.2045;
    /* Note DD.MMSS Global coord. format */

    (void) printf("Entity type %s Included term %s with
        attributes %s is located at %f %f\n",
        Feature.entity_type, Feature.entity_attributes,
        Feature.included_terms, Feature.Point.x,
        Feature.Point.y);
    return;
}
```

The declaration and assignments shown in Function 7.7a would store this information. Similarly, the same church may have a specific name to be placed next to a symbol for a church on a map, such as "St. Paul's Church". In this case, a label point should be used (Function 7.7b).

Function 7.7b

```
/* Program segment to demonstrate a label point
/* structure kcc 4-94 Revised from 12-88 funct 7.7b */
#include "cart_obj.h"
(void) show_label_point()
{

   LABEL_POINT Feature;
   Feature.label = "St. Paul's Church";

   Feature.Point.x =  (coordinate) 1.3000;
   Feature.Point.y = (coordinate) 55.2045;
   /* Note DD.MMSS Global coord. format */

   (void) printf("Label Point: Label %s is located \
     at %f %f\n",
      Feature.label, Feature.point.x, Feature.point.y);
   return;
}
```

In the case of the area point, the identifier is a pointer to a polygon which is associated with the point. For example, the pointer could be to a polygon defining the boundaries of the church parish. Similarly, the node simply defines the location and gives an identifier so that the node can be found by other, higher-dimensional objects. The first encounter within a computer program in the C language with these structures is when the data are read from files. We will use ASCII files throughout this discussion, although very much faster input and output to files is possible if binary reads and writes are used. As noted above, ASCII data files can be edited, examined, modified, and corrected with the use of your favorite text editor. A function to read a single file containing a large number of

---

Function 7.8

```
File:    point_data
Format:  ASCII


1.3000      55.0000     St. Paul's Church
2.1023      56.1012     Water Tower
-1.1213     57.1123     Bench Mark (45.4 meters)
-1.1557     55.5927     Radio Tower
2.1519      54.1314     Radar Reflector
```

---

point features with their labels follows. Assume that the file called `point_data` as shown in Function 7.8 exists in the users current directory. Function 7.9 will load the data into the structure `label_point`. This function, `read_label_points()`, should be declared in the calling function to be of type `int`. The function returns the number of points read from the file via the `return` statement, plus the structure containing the data through the argument list. The calling function should provide in the argument list the address of a structure ready to receive the data.

Function 7.9

```
/* Read a file of latitudes and longitudes in DDD.MMSS
/* format with point labels and load the data
/* into a structure of label points
/* kcc        4-94 revised from 12-88 funct 7.9 */
#include <stdio.h>
#include "cart_obj.h"
int read_label_points(label_point)
     LABEL_POINT Label_point[];
     {
     int i=0, number_of_points;
     FILE *file_pointer;
     /* Open the input file and abort if non-existing */
     if ((file_pointer = fopen("point_data","r")) == NULL)
          { printf("Unable to open file 'point_data'\n");
          exit(); }
     /* Read the point data, counting entries */
     while (fscanf(file_pointer,"%f%f%s",
          &Label_point[i]->Point->x,
          &Label_point[i]->Point->y,
          Label_point[i]->label) != EOF) i++;
     number_of_points = i;
     /* Close the file */ fclose (file_pointer);
     /* Return the number of points */
     return (number_of_points);
}
```

7.3 ONE-DIMENSIONAL CARTOGRAPHIC OBJECTS

The one-dimensional cartographic objects are more varied, serving a large number of cartographic needs (Figure 7.2). The generic name for a one-dimensional object is a *line*, which can consist either of a locus of points defined by a function such as a polynomial or a b-spline (an *arc*) or as a sequence of connected primitive objects known as line segments. The *line segment* is simply a straight line connecting two points, and we define it to be simply two points stored together.

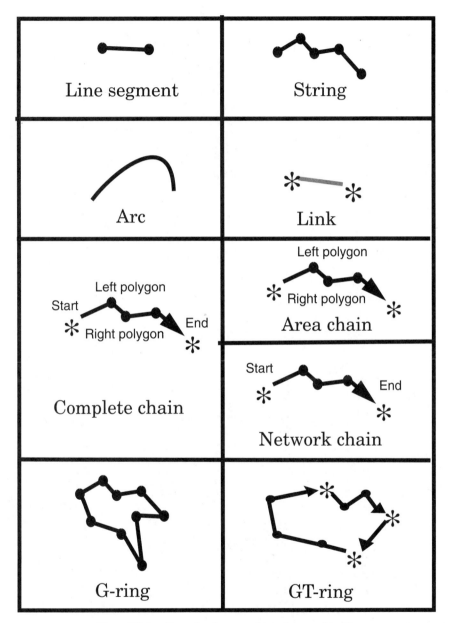

Figure 7.2 One-dimensional cartographic and geographic objects.

Function 7.10

```
/* ------------------------------------------
/* C language cartographic data structures
/* Based on FIPS 173 Terminology and data model
/* kcc 6-93 Revised from 5-88 function 5.10 (7.7 Ctd.)
/* ---------------------------------------- */
/* One-dimensional objects */
typedef struct {
     POINT Point[2]; } LINE_SEGMENT;
typedef struct {
     int number_of_points;
     POINT Point[MAXPTS]; } STRING ;
typedef struct {
     int arc_identifier;
     STRING String;
     int *function; }  ARC;
typedef struct {
     NODE Node[2];} LINK;
typedef struct {
     NODE Node[2];
     int left_polygon, right_polygon;
     STRING Chain;} COMPLETE_CHAIN;
typedef struct {
     NODE start;
     NODE end;
     int left_polygon, right polygon;
     STRING Chain;} AREA_CHAIN;
typedef struct {
     NODE start;
     NODE end;
     STRING Chain;}  NETWORK_CHAIN;
typedef struct {
     int number_of_strings, number_of_arcs;
     STRING String[MAX_STRING];
     ARC Arc[MAX_ARC];}  G_RING;
typedef struct {
     int number_of_chains;
     COMPLETE_CHAIN Chain[MAX_CHAIN];} GT_RING;
typedef struct {
     STRING Ring;} RING;
```

The *string,* a group of connected segments, without any other topological information, makes up the building block for the more complex one-dimensional objects. When a string closes upon itself, it is termed a *ring,* and rings can be created from strings, arcs, links, or chains. The *link* is a network primitive, consisting of a single edge of a connected graph. Direction within a link is simply by the sequence of points as stored.

This is a complex group of objects, and there is a need to break the list down. A simple division is by those objects that store topology, and those which are designed largely simply to reflect geometry. Traditionally, cartography has been concerned with using geometry-only objects to represent cartographic data which is destined solely for display. Nevertheless, topology is often required for consistency checking, sequenced color fills, and symbolization. Topology is also required for some data structure conversions. The geometric objects of one dimension are the line segment, the string, and rings of arcs and strings.

To read a string, assume a file containing data in the format of one string per line, with the first record being the number of points in the string (n), and the next records being the coordinate pairs, as (x1,y1), (x2,y2), (x3,y3) up to (xn,yn). Such a file, containing the decimal latitudes and longitudes of the outline of North America, is contained on the companion disk in the file namerica. Similarly, a G-ring is identical to the string except that the first and last point *x* and *y* values match exactly and that the string may contain arcs.

Function 7.11 shows the two functions that perform the task of reading a string. The function get_pts() reads a single string into a structure of type STRING and assumes that the header file cart_obj.h both defines STRING and declares MAXSTRINGS. Here inclusion comes by being in the same file. The actual structure containing each string, called line, is declared here outside the brace defining the function. The function read_strings makes successive function calls to get_pts(), each time storing an additional STRING structure into line. When the function finds the end_of_file marker, the function read_strings() returns the number of successfully stored lines in the structure.

The same function will work for a G-ring, but a check should be implemented to ensure closure and to ensure that no arcs are involved. The primary purpose for a G-ring is for either the computation of an area or the coloring, shading, or filling of the area delimited by the G-ring with a pattern.

To read an arc, the user should first read an identifier for the arc, next read a string in the same way as in function get_pts(), and finally, read a pointer to a function. The user must determine how the function's parameters are stored. For example, if we are limited to circular arcs, we could store the (x,y) of the circle center, the circle radius, followed by the starting and ending circular arc bearing, clockwise from north, and the number of points to be generated along the arc. Thus four floating point numbers, one coordinate pair, and one integer would be needed to define the arc. The STRING part of the structure can then hold the points that are generated, at any given increment.

Function 7.11

```c
/*=============================================
* Read a string from a file into a structure
* kcc 6-93 funct 7.11
/*========================================= */
#include <stdio.h>
#include "cart_obj.h"
#define INPUT_FILE "namerica"
/* read_strings : reads strings into structure */
STRING Line[MAXSTRINGS];
int read_strings()
{
    STRING get_pts();
    int i, j, number_of_lines = 0;
    FILE *fp;
    fp = fopen(INPUT_FILE, "r");
    while (!feof(fp)) {
        Line[number_of_lines++] = get_pts(fp);};
    number_of_lines--;
    (void) printf("%d lines read from %s\n",
        number_of_lines, INPUT_FILE);
    return (number_of_lines);
}


/* get_pts : reads points for one string */
STRING get_pts(fp)
    FILE *fp;
    {
    STRING get_line;
    int i, j;
    (void) fscanf(fp, "%d%*1c",&get_line.number_of_points);
    for (i = 0, j = 1; i < get_line.number_of_points; i++) {
        fscanf(fp, "%f%f", &get_line.Point[i].y,
        &get_line.Point[i].x);
        if (j++ == 6) {j = 1;(void) fscanf(fp, "%*1c");}
    }
    if (j != 1) fscanf(fp, "%*1c");
    return (get_line);
}
```

The RING structure in Function 7.10 is not explicitly part of the SDTS definitions, but is included here because by far the majority of simple applications use G_RINGs consisting of only one string to define a topology-less polygon. Storing multiple strings and arcs for every instance of a G_RING would use an unnecessarily large amount of storage in these cases. The burden of ensuring closure is then left to the user. Closure in the RING can be implicit (that is, the software will duplicate the first point as the last) or required.

The remainder of the one-dimensional objects are designed to store both topology and geometry. These are the link, the complete chain (called simply a chain here), the area chain, the network chain, and the GT-ring. The latter objects are included as components of aggregate spatial objects, considered in Section 7.4. The simplest of these is the link.

A link consists simply of two nodes, stored here as a line segment with an implied ordering by the order of storage. A very common application of the link is to show traffic flow. A traffic flow network in Manhattan has a series of links, because each street from intersection to intersection is a straight line segment. Where streets curve, as in Central Park, the network chain would replace the link.

Taken together, the whole street network would be described by the network, one of the aggregate spatial objects defined in the Spatial Data Transfer Standards. For such a network, the geometry is important for computing lengths and drawing the map, but much of the topology can be computed simply from the endpoints of the network chains stored separately as links.

The various versions of the chain, other than the network chain, are usually encountered in applications that plot chain by chain rather than string by string or polygon by polygon. Although it is possible to check for topological continuity using the chain structures, their format is inherently one-dimensional; that is, it says little about the areas enclosed by the chains.

In the case of the GT-ring, it is known that the chains connect together perfectly to produce a ring, a property that differs from the G-ring because a string component of a G-ring does not have to be free of crossovers or intersections by definition. This sort of distinction is often important in digitizing, where endpoints are snapped together to maintain connectivity along chains. Usually, the digitizer operator must separately signal a chain beginning and end, perhaps by pressing different cursor buttons, and must sometimes enter connecting chain information from the keyboard.

7.4 TWO-DIMENSIONAL CARTOGRAPHIC OBJECTS

Two-dimensional objects under the final version of the Spatial Data Transfer Standards differ to the greatest extent from the draft version of the standard in 1988. An *area* is defined as a bounded, continuous, two-dimensional object that may or may not include its boundary. The defined two-dimensional objects are the *interior area,* the *G-polygon,* the *GT-polygon,* two special case polygons (*void* and *universe*), and finally the raster objects the *pixel,* and the *grid cell* (Figure 7.3).

The interior area is an area excluding its boundary. This area is filled when a polygon is colored or shaded. It is also the area within which some algorithms, such as point-in-polygon tests, provide a simple solution. The G-polygon is defined as an area consisting of an interior area, one outer ring, and zero or more nonnested inner rings. The ring

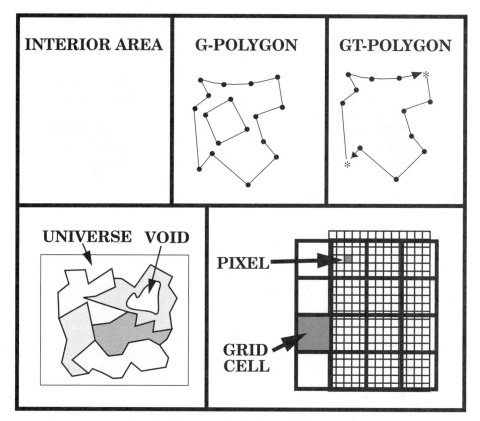

Figure 7.3 Two-dimensional cartographic and geographic objects.

is the standard representation of the polygon for shading, area computation, and filling in most cartographic applications.

The GT-polygon is similar to the G-polygon, with the exception that the boundary is topologically defined as a GT-ring built from chains rather than simply a G-ring. Two degenerate forms of the polygon are the *universe* polygon and the *void* polygon. The universe polygon makes up the area within the bounding area of the map's extent, but it is outside the area for which attributes and objects are defined. On maps, this area is often bounding countries, states, oceans, and so forth. The void polygon is an area without attributes within the figure of the map. Thus a map of North America may consider the Great Lakes as void polygons, especially if the map depicts a land-related variable such as population density or land use.

The two final two-dimensional cartographic objects accord well with the GKS standard for graphics and are designed to facilitate the handling of raster and grid data. The *pixel* is a picture element consisting of the smallest nondivisible element of an aggregate object called a digital image. The grid cell is the final object and is an element of a regular or nearly regular tessellation of a surface, an aggregate object called a grid. Thus for DEM data, for example, the grid cell is an elevation observation spaced 30 meters apart

Function 7.12

```
/* ---------------------------------------------
/* C language cartographic data structures
/* Based on FIPS 173 Terminology and data model
/* kcc 6-93 Revised from 5-88
/* function 7.12 (Continues 7.10)
/* --------------------------------------------- */
typedef struct {
      G_RING interior;
} INTERIOR_AREA;
typedef struct {
      int polygon_identifier;
      char polygon_descriptor[MAXLABEL];
      struct G_RING Boundary;
      int number_of_holes;
      struct G_RING Hole[MAXHOLES];
} G_POLYGON;
typedef struct {
      int polygon_identifier;
      char polygon_descriptor[MAXLABEL];
      struct GT_RING Boundary[MAXCHAIN];
} GT_POLYGON;
typedef struct {
      char grid_descriptor[MAXLABEL];
      int nrows, ncols;
      POINT corners[4];
      int datum;
      int z[MAXROW][MAXCOL];
} GRID;
typedef struct {
      int nrows, ncols;
      unsigned char *pixels; /* Assumes 8 bit data */
} IMAGE;
```

from its neighbors on a square tessellation, while multiple pixels could be in use in a digital map to symbolize the grid cell. The two-dimensional objects can be represented by the C language programming data structures shown in Function 7.12, which continues the definition of the `cart_obj.h` header file mentioned previously.

The interior area is represented as a G-ring. Algorithms for computing area and testing a point for inclusion within an area or on the boundary are considered in Chapter 10.

The G-polygon structure represents a normal nontopologically constructed polygon. This structure contains an identifier, a descriptor (such as the label "New York State"), and a single structure of type G_ring to contain the boundary. In addition, the number of holes must be stored, and the G_RING structure is multidimensional to store the holes.

A maximum number of holes must be specified, stored in cart_obj.h as MAX-HOLES. We could use C's pointer capabilities to store holes, that is, to store within the structure a pointer to a structure of type G_RING that constitutes a hole. Although such a structure is elegant and allows complex polygons to contain any number of holes without wasting storage, this addition will be left to the more advanced C programmer. The pixel, being largely device dependent, is excluded from the list of structures. This is because the GKS standard provides a means for symbolizing a grid without reference to pixels.

The remaining structure is the *grid*. The grid has as its elements a grid descriptor (which could be the header lines from the USGS DEM data files), POINT references to the ground map coordinates of the four grid corners (which could, for example, be UTM values) and a two-dimensional array of integers to store the values. Elevations on land range from 400 meters below sea level on the shore of the Dead Sea to 8,846 meters at the summit of Mount Everest.

In case the range of values required does not fit into the range of an integer, the datum value is provided. This integer value should be added to the grid value to give the actual elevation. For most DEM data, this is zero, but for ocean depths, elevations on other planets, and other geographic variables, other values may be necessary.

Although the grid data structure is a logical construct for the grid, for practicality, the data in a grid are structured independently of the map reference information. Variants on the grid structure that conserve the significant amount of storage necessary when an array is allocated in computer memory are possible. When a grid data set is detached from its spatial frame of reference, it can be called an *image,* a term not used in the Spatial Data Transfer Standards.

An image can be of varying size and needs to be able to store different values in each pixel (note that grids have grid cells; images have pixels). An image with a given type can have its storage size precomputed, and then allocation can be made only for the space it requires. The data structure here is borrowed from Myler and Weeks (1993) to which the interested reader is referred for a cornucopia of image-based algorithms, including the reading and writing of several industry-standard image formats and basic and advanced image-processing algorithms.

The value of using the structure of an image is that allocation of image space can be made dynamically within a program, and that types of different sizes can be accommodated. The type of the pixel need not always be an integer, which often uses far more storage than is necessary. Common types are short, unsigned int and unsigned char. For example, a grid's elevations can be allocated to an image in the following way (Function 7.13).

Function 7.13

```
/* Function grid_to_image: Allocate a set of grid cells
   to an image function 7.13 kcc 4-94 */
IMAGE grid_to_image (grid)
GRID grid;
{
    IMAGE *image;
    int i, j;
    if (image = malloc (2 * sizeof (int) +
        (grid.nrows * grid.ncols *
        sizeof (*image->pixels))) == NULL) {
    printf ("Unable to allocate memory in function \
        grid_to_image()\n;
        return(1)
    }
/* Image allocated dynamically, now copy grid to image */
    image->nrows = grid.nrows;
    image->ncols = grid.ncols;
    for (i=0; i< grid.nrows;i++ ) {
        for (j=0; j<grid.ncols;j++)
        *(image->pixels++) = grid.z[i][j];
    }
    return (image);
}
```

7.5 DATA STORAGE AND DATA STRUCTURES

At least one major problem with the data standards is the lack of three-dimensional cartographic objects. Fortunately, a small set of cartographic data structures for three-dimensional objects exists, covered in Chapter 13. In CADD, and in computer graphics generally, data structured for three-dimensional objects are a vital part of their symbolization or rendering as pictures. The GKS graphics standard has been extended to three-dimensional graphics (GKS-3D), which may tend to favor particular three-dimensional data structures in the future.

The C language structures introduced in this chapter are used extensively in the chapters that follow. In Chapter 8, they are used to give examples of the data structures in common use, while in Chapters 10, 11, and 12 these basic programming structures are built upon with transformational algorithms and symbolization methods using GKS. Although their understanding is not critical to students of computer cartography, students of analytical cartography are strongly encouraged to use them to experiment with their own software for computer and analytical cartography.

7.6 REFERENCES

Department of Commerce (1992). *Spatial Data Transfer Standard.* FIPS PUB 173, Federal Information Processing Standards Publication, Computer Systems Laboratory, National Institute of Standards and Technology, Gaithersburg, MD.

Guptill, S. C. ed. (1990). *An Enhanced Digital Line Graph Design.* Department of the Interior, U.S. Geological Survey Circular 1048. Washington, DC: U.S. Government Printing Office.

Myler, H. R. and A. R. Weeks (1993). *The Pocket Handbook of Image Processing Algorithms in C.* Englewood Cliffs, NJ: Prentice Hall.

National Committee for Digital Cartographic Data Standards (1988). "The Proposed Standard for Digital Cartographic Data. " *American Cartographer,* vol. 15, no. 1, pp. 9–142.

8

Map Data Structures

8.1 WHY MAP DATA STRUCTURES ARE DIFFERENT

A distinction can be made between map and attribute data structures. Map data structures are different because they store information about location, scale, dimension, and other geographic properties. Attribute data structures may be linked to the data structures for cartographic objects, but they contain attribute information about the objects, or links between the objects only. In this chapter, we consider only the data structures dealing explicitly with locative information. In Chapter 9, we focus on attribute data structures. Map data structures are particularly cartographic and have been developed by cartographers and others to deal with computer and analytical needs and for use in GISs. Attribute data structures can also be called data models and are used to organize text or numerical data as well as cartographic data. Also essential are how these two types of data structures are related to each other, and how data can be transformed between structures. This theme is the topic of Chapter 12.

In the early days of computer cartography, the data sets that went along with cartographic software contained the minimum amount of information necessary to produce the desired map and as such used data structures that were nonanalytic, device specific, and as a result short-lived. These data structures, which have become known as entity-by-entity, or "cartographic spaghetti, " structures remain in circulation and for many purposes still serve the needs for which they were intended. Early on, however, attention became focused on the potential power of topologically encoded cartographic data. A group at Harvard University, at the time producing the ODYSSEY package, sponsored a conference on topological data structures that became a milestone in the acceptance of the need for topological encoding (Dutton, 1979). These topological structures have become the data structures for a large number of data sets such as DLG, GIRAS, and TIGER, and they form the core of a majority of the systems for the display and analysis of cartographic data.

In 1975, Tom Poiker (formerly Peucker) and Nick Chrisman published a paper enti-tled "Cartographic Data Structures", that reported a way in which the information re-quired for the topological encoding of cartographic data could be converted into a data structure that was particularly geographic. Their structure, part of a program called POLYVRT, became the model for many cartographic systems and was highly influential in the evolution of data structures (Peucker and Chrisman, 1975). Since then numerous new, efficient, and thought-provoking data structures have been devised for cartographic data. The grid structure, the quad tree, and tessellations, in addition to the entity-by-enti-ty, and the topological structures, have found acceptance and use within computer map-ping systems.

A *data structure* can be defined as a logical organization of information to preserve its integrity and facilitate its use. By *information* here we mean the information content inherent in cartographic data, especially that necessary to produce a map. The informa-tion relates to the fundamental properties of geographic objects, that is, information about size, location, shape, distribution and so forth. We have to *preserve the integrity* of this information. In other words, we have to store it with correct precision, in a coherent and consistent manner, and in such a way that we can retrieve the right information at the right time. To *facilitate the use* of the information we need to have some way of organizing our data physically on a storage medium in a method of representation that allows us to sym-bolize cartographic objects on maps with relative ease and within a reasonable length of time. Much work in computer cartography consists of writing data into cartographic data structures and transforming the data between structures. Different structures suit different kinds of mapping and different sets of demands.

The general characteristics of cartographic data structures are largely input deter-mined. The data structures inherent in the geocoded data are usually determined by what-ever input device was used to capture the data. Whether it was a digitizing tablet or an orbiting satellite, the structure that the data collection instrument imposed on the data is almost always the form in which we get the data. As the digital cartographic data stan-dards gain acceptance, most existing data will be in more logical formats, but for the present, the input-determined nature of the data is a fact.

In many cartographic applications, the data need only be a simple representation of the map in order to produce a graphic. Integrating data sets, such as plotting the various separations that make up a multicolor topographic map, simply involves overlaying one data layer onto another. The cartographer need only decide upon an order of precedence. In additive color systems, the last color plotted at one location takes priority, and no mix-ing takes place, although there are exceptions. In these types of data sets we have no ex-plicit relations between the cartographic objects. The only geographic property represented is location, and the map interpreter is left the task of searching for geographic relationships.

Thus we have two types of data structure in analytical and computer cartography. First, we have data structures to store strictly locational data about the map, with perhaps any necessary topological or other data required to produce the map. In addition, we have attribute data structures, which encode the attributes associated with cartographic objects, plus the indexing necessary to support query and retrieval functions on the objects. A map data structure is the minimum required for a computer mapping system, plus the actual data with its representational characteristics, and its origins in geocoding.

We can now respond to the issue raised at the beginning of this section. Why are map data structures different? Map data structures are required for computer mapping, and they are different from attribute data structures in that their purpose is to support computer cartography and not necessarily analytical cartography. A map plus an attribute data structure is the minimum requirement for the additional analytical functions we may use in analytical cartography, in the more sophisticated displays of advanced computer mapping systems, or in GISs. Also, many of these systems owe their power and capabilities to how the map and attribute data structures are related to each other, both logically and physically. Such maps—with both map and attribute, or graphic and non-map, data fully integrated—are often called "smart maps."

8.2 VECTOR AND RASTER TECHNOLOGIES AND DATA STRUCTURES

We have already seen a hardware division between raster and vector systems. Just as there are certain input controls on data structures, there are certain output constraints on data structures. Both input and display technologies have reflected generally a movement from vector-based to raster-based over time. Vector input devices such as digitizing tablets and output devices such as pen plotters are increasingly being replaced with raster technology, such as scanners for input and raster displays for output. This has influenced thinking about data structures considerably.

Both raster and vector data structures have been proposed as the answer for structuring geographic data. To use either one to the exclusion of the other, however, means accepting the inherent disadvantages of one of the systems. Many recent computer mapping systems and GISs have solved this problem by supporting both structures and allowing the user to transform between structures as appropriate. Peuquet (1979) showed that most algorithms using a vector data structure have an equivalent raster-based algorithm, in many cases more computationally efficient. Logically, therefore, the user should structure data according to the relative merits of each data structure for a particular type of cartographic entity or a particular mapping circumstance. Burrough (1986) has provided an extensive list of the advantages and disadvantages of each type of data structure. The advantages and disadvantages are worth considering, because they reflect the correct choice of structure for dealing with particular types of data and particular geographic properties.

8.2.1 Advantages of Vector Data Structures

Burrough considered vectors a good representation of data mapped as lines and polygons, because vectors can follow the lines very closely. This means that maps produced from vector-based mapping systems are capable of greater resolution and accuracy. This, however, assumes that the lines that have been captured from the map as vectors are thin, and line thickness at the map scale is not an issue. The vector structure is also compact in that it can represent very complex lines with a minimal amount of information. We have seen that in vector mode we can reduce redundancy because if we have a fairly long line segment we only need to represent it with two x, y pairs, the two endpoints. Many other systems embed redundancy in representing straight lines.

As far as topological data structures are concerned, vectors are preferred, because topology can be completely described within vector-based systems as network linkages. Vector structures support interactive retrieval, so that updating and generalization of map data and data attributes are possible. For example, if we have a vector representation of a line at 1:25,000, it is very easy to resample the points along that line to give a generalization of that line so that we can plot the map at another scale, such as 1:250,000. Also, in editing the map we can remove an entire chain and reenter it if we have made an error.

8.2.2 Disadvantages of Vector Data Structures

Vector data structures are complex and are less intuitively understood by the users of cartographic data. The biggest disadvantage, however, is that the combination by overlay of two or more vector-based maps is a very computationally intensive task. Map overlay, or computation of the most common geographic units, means processing networks to find where they intersect. Just finding the intersections is difficult because we have to test every line segment of every chain to find an intersection unless we use some kind of preprocessing based on the ranges in x and y of the chains. We next have to reorder the polygons by the new intersections.

Overlay is a classic geographic problem, and many geographic phenomena need to be redistributed between sets of irregular areas. Not the least important, for example, is the redrawing of congressional and voting districts in the United States after each decennial census.

Display and plotting of cartographic data in vector mode can be expensive, particularly higher-quality color and cross-hatching. A vector-mode device especially does not do area fills very well, because plot devices try to fill areas by dragging the pen backward and forward, a time-consuming process. Vector mode display technology is also expensive, particularly for the more sophisticated software and hardware.

8.2.3 Advantages of Raster Data Structures

The principal advantage of raster data structures is their simplicity. The grid is an integral part of cartography and has long been used to structure geographic information. More recently, cartographic data have been acquired in grid formats directly, for example, in satellite and scanned air photo data. This aspect is becoming increasingly important as remote sensing becomes a source of new cartographic data.

Using the grid structure, analytical operations are easier, such as computation of variograms, and some autocorrelation statistics, interpolation, and filtering. Resolution on raster devices is improving steadily, and the cost of the graphic memory boards that support display in raster mode is falling, whereas vector technology has not changed significantly in cost or capability since the mid-1980s.

8.2.4 Disadvantages of Raster Data Structures

The principal drawback to raster data structures is the required volume of data. Grids embed much redundancy, so have a much larger data volume. If we have extremely variable data, such as topography, then grids use storage inefficiently, because compression methods do not save significant amounts of space. Also, the use of large cells to reduce data volumes means that some cartographic entities can be lost, simply slipping through the sampling net. Raster maps are considerably less visually effective than maps drawn with lines. Text on low-resolution raster devices is often illegible. Also, curved and angled text is difficult to produce. In raster mode, network linkages are difficult to establish. Cartographic entities such as stream networks, or boundaries between polygons, or linkages between town centers, are not well supported as cartographic objects. It is difficult, for example, to establish links within a stream network in raster mode. Map projection transformations are time-consuming using raster data structures, unless special algorithms or hardware are used. A disadvantage often ignored is orientation. If we use a grid, we imply an orientation to a grid, which automatically means that the sampling strategy has a particular directional bias. We normally orient our grids north/south and so bias the data against other directions.

As we have seen, neither of the data structures holds all solutions for all data. Grids are usually used for mapping extensive geographic information at small scales, while vectors are used for detailed information at large scales. Systems that support both structures allow data to be structured to take advantage of the strengths of each structure type.

8.3 ENTITY-BY-ENTITY DATA STRUCTURES

Cartographic entities are usually classified by dimension into point features, line features, and area features. The simplest means of digitally representing cartographic entities as objects is to use the feature itself as the lowest common denominator. Using entity-by-entity data structures, we are concerned with discrete sets of connected numbers that represent an object in its entirety, not as the combination of features of lesser dimension.

The simplest entity-by-entity structure is the point. In the digital cartographic data standards, the cartographic objects listed as "graphics only" are closest to entity-by-entity structures in type. Thus the point, especially the entity point, is the simplest form for dimension 0; the string, the arc, and the ring, for dimension 1; and the polygon and grid cell for dimension 2.

Chapter 6 included program segments for reading and storing data in these structures, and Chapter 11 contains the necessary code for generating the map from the data in these structures. The grid cell structure is covered in detail later in this chapter.

An important difference between entity-by-entity and topological structures is shown by the ring structure. As far as a map is concerned, a lake can be shown on a map using a RING data storage structure. The first and last points match perfectly in x and y is simply to close the polygon visually. If we wish to compute the lake area, or the length of the boundary, or even shade the lake with color, the ring structure is adequate. If we wish to find the polygons that border the lake, or determine which rivers flow into the lake, however, it would require a large amount of systematic searching to find endpoint matches.

Similarly, for a point, we are able simply to plot a symbol at the point, but we do not know which other features are related to the point and how.

8.3.1 Point Objects

Attributes associated with cartographic objects stored in entity-by-entity data structures rarely go beyond those contained in the C language structures in Chapter 6. Point features need only a feature code, because the feature code can be used as a key into attribute database of independent records for each feature. An entity-by-entity list of cities could contain their locations stored as a set of points, and if populations were required to produce a proportional circle map, for example, the cities could be stored as label points and the label itself could be used as an index into a file of attributes. The files in Function 8.1 could contain all the data necessary to produce a proportional symbol map.

Function 8.1

```
File : City_Populations
Format : ASCII population est. in 1986
Source : Goode's World Atlas, 17th Ed.
Structure : Flat File
```

Quito	918884
Rabat	367620
Rangoon	2276000
Rawalpindi	452000
Recife	1204738
Riga	875000

```
File : City_Points
Format : ASCII Lat/Long in DDD.MM format
Structure : Label Point
```

Quito	-0.17	-78.32
Rabat	33.59	-6.47
Rangoon	16.46	96.09
Rawalpindi	33.40	73.10
Recife	-8.09	-34.59
Riga	57.56	23.05

8.3.2 Line Objects

In an entity-by-entity structure, a line is usually represented by a string, that is, an ordered set of points, which connected together in sequence trace out the line. In vector mode the string is unconstrained, although usually lines are terminated and restarted when crossing takes place. Similarly, there is no obstacle to double specification of a single line, a common problem resulting from the manual digitizing process. A line can carry an attribute, such as a traffic volume, a color, a thickness, or a feature type such as highway or railroad. Attributes are not really assignable to strings directly, but a separate file of attributes by string number can be maintained. Alternatively, separate files for each feature type can be maintained, similar to separating line colors by separation in a manual color production process.

A common alternative to keeping a file containing each string listed point by point is to use a point dictionary (Figure 8.1). In this file arrangement, separate files are kept for points and for lines. The point file is simply a listing of every point used on the map, containing an identifier, often in a sequential order, and an easting and northing. The second file, the line file, contains line numbers, perhaps the line attributes, and the starting and ending point numbers of sequential groups of points in the points file that make up the line. The point numbers act as pointers into the point dictionary. An alternative system has a list in the lines file of every point identifier in the line. This system, although much more cumbersome to implement and use, has the advantage that the point data file can be sorted at will to facilitate searching or processing. Point dictionary systems make editing and data entry relatively difficult, but have found their uses in mapping systems.

Several grid-type structures exist for entity-by-entity data structures for lines. If a grid is used, lines are typically labeled with attribute numbers that are assigned to grid cells

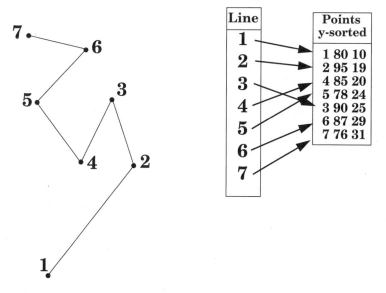

Figure 8.1 Line defined as entity-by-entity reference to point dictionary.

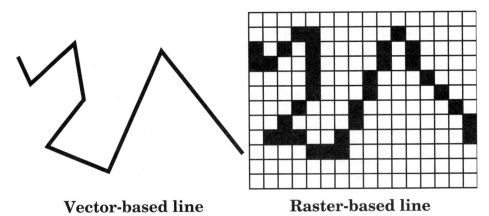

Vector-based line   **Raster-based line**

Figure 8.2 Representation of lines in entity-by-entity structures.

as data. Thus to give an entity-by-entity definition of a river for use in a grid mapping system, we could assign a value for the river to every pixel into which the river passes. Connectivity can be limited to the four adjacent cells or enlarged to permit diagonal or eight-cell connectivity. It is normal to thin the lines to one pixel width for processing and display (Figure 8.2). If lines are stored as attributes or index numbers in the grid, problems result at line intersections. These are sometimes stored separately; otherwise, the attribute or color plotted at an intersection is the last one referenced.

The Freeman code is an entity-by-entity line data structure that falls between vector and grid. The Freeman code is a representation of a line as a sequence of numerical codes, each representing the direction of a step moved along the line (Figure 8.3). Freeman codes that use one octal digit per code are eight-pointed, while one hexadecimal digit (nibble) per code allows 16 values. A problem of converting grid data to Freeman codes is that diagonals have length of $\sqrt{2}$ over 2, wheras the primary directions have length of one over two in grid units. Both strategies can be used. Anstaett and Moellering (1985) gave a simple algorithm for computing areas of polygons bounded by sets of Freeman codes. Freeman codes can also be run-length encoded, meaning that files containing long straight lines, with repeating identical Freeman codes, can be compressed considerably.

8.3.3 Area Objects

Because we are dealing with entity-by-entity structures, polygons of two-dimensions can be represented cartographically by one-dimensional objects unless we wish to fill the polygon. Appropriate structures in the C language, as given in Chapter 6, are the two forms of the RING, in its form as a set of strings or as a set of arcs. As a two-dimensional object, the POLYGON structure is appropriate. Strictly speaking, the complex polygon with holes cannot be stored as an entity-by-entity object. The hole can be part of the boundary, with a "bridge" between the two lines, or we can use alternative means to include the hole.

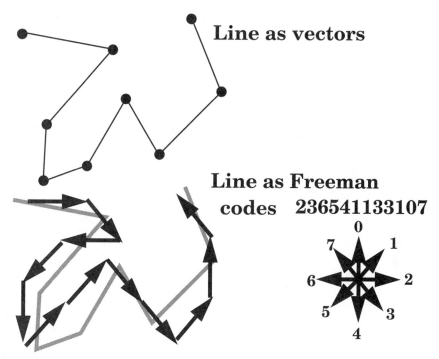

Figure 8.3 Freeman codes as a line data structure.

Just as we used the point dictionary for a line, we can also use the point dictionary for an entity-by-entity area (Figure 8.4). In this case the second file is the area file, which contains the polygon identifier and a list of the points that make up the boundary. There may be a good reason for using this structure, because a sorting of the point dictionary makes it simple to determine shading instructions for filling the polygon. A polygon list, in addition, is the simplest means for computing the area of a polygon and for drawing the polygon as a single entity against a plain background on a map.

Using the grid structure, polygon representation is simple. Membership is usually assigned to the grid cell with the highest area content for any polygon, and all cells within a polygon are assigned one value, either an index or an attribute (Figure 8.5)

Alternatively, the class of the center of the grid cell can be assigned. Holes can simply be assigned last, so that their attributes fall on top of any prior values. In this way, areas can be computed by cell counting, a rapid though less accurate technique. Similarly, polygons can be made up of strings of Freeman codes, although in this case care must be taken to assure closure.

Closure is when the vector sum of the individual Freeman code vectors is zero, but again no requirement to prevent crossing is included. Crossed-over polygons (Figure 8.6) are sometimes called weird or splintered polygons. These are often errors in digitized maps that we seek to detect and eliminate. Area calculations and other operations are not appropriate for these polygons.

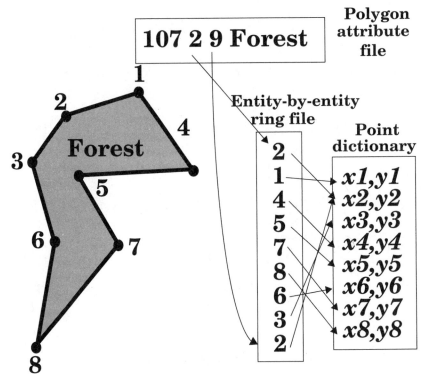

Figure 8.4 Point dictionaries for a polygon.

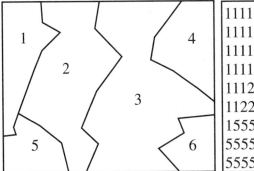

Figure 8.5 Polygon structures for a grid.

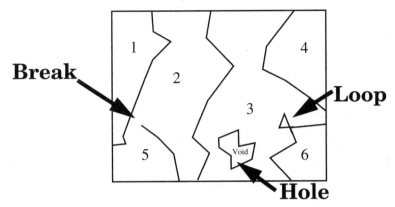

Figure 8.6 Weird polygons.

8.4 TOPOLOGICAL DATA STRUCTURES

Entity-by-entity data structures, although useful for the symbolization of cartographic objects, are often unsuitable for more complex analysis because they store no information about topology. Furthermore, we have no guarantee with entity-by-entity structures that impossible geometric configurations such as weird polygons and slivers do not exist. Topological data structures store the additional characteristics of connectivity and adjacency. As such, they allow immediate checking, perhaps during data entry, for errors in geocoding and data storage.

The basic objects are the node, the link, and the chain. To these are added direction, as in the link, and higher structure, as in the network chain. Nodes are points with topological significance and as such must be labeled. The link and the network chain are similar in that they support linkages to adjacent objects and nodes. Connectivity to nodes is essential to give a graphic frame of reference and to link lines in the network together. The direction and the links to other primitive objects can be summarized as properties of a winged segment (Figure 8.7).

Linkages are sometimes stored as forward linkages and reverse linkages. This is because chains can begin only at one end, and to piece together objects it is very often required to traverse a chain in either direction. Forward linkages are the segment identifiers of connected segments that are pointed to when a chain is traversed in the direction in which its x and y values are stored.

We can have forward linkages to other links, directed links, or chains. Usually, there is a finite limit to the number of chains that can meet at a node. This is not a problem for most maps; for example, the 48 contiguous states have only one node where four lines meet (at the Four Corners), but sometimes at the poles, for example, many lines meet together at once.

To traverse an entire network we only need know immediate right-hand and left-hand turns. Eventually, if we keep following around the network, we come back to the start point along some other chain. If we have made an error and one of the chains has not been

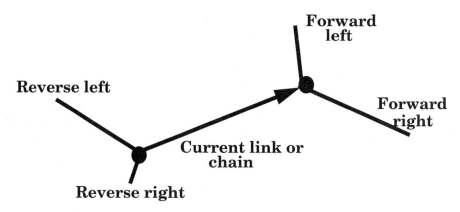

Figure 8.7 A winged segment.

snapped to a node, or if not all intersections are nodes, the network tracking algorithm in trying to close the polygon will go off along the wrong chains, going around through the network trying to close a polygon.

Topological data structures also store polygon information, which makes the data structure more analytically powerful. Right-hand and a left-hand polygons are assigned according to the direction of the chain or directed link. This allows us to presort the chains to select all chains that say that they are neighbors of a particular polygon, for example. These chains can then be connected either by their chain linkage information or by matching their end points, and they can be converted into a polygon list as an entity-by-entity structure or simply plotted in order.

8.5 TESSELLATIONS AND THE TIN

Tessellations are connected networks that partition space into a set of sub-areas. They consist of the network itself and the spaces left within the areas. Special cases of tessellations are topological structures, in which the space is partitioned into regions of geographic interest such as states and counties and grids, in which the tessellation is a network of connected squares and the grid cells they divide the space into. The remainder of tessellations include a number of systems used infrequently in analytical and computer cartography, such as equilateral triangles and hexagons, which are geometric and regular in nature, and one irregular system, which has gained widespread use, the *triangulated irregular network* (TIN).

Mark (1975) has suggested that TINs are more accurate and use less space than any other data structure for topography, and McCullagh and Ross (1980) demonstrated that TINs can be generated from point data faster than the alternative data structure for terrain, the grid. TIN data structures can describe more complex surfaces than a grid, including vertical drops and irregular boundaries. Because single points can be easily added, deleted, or moved, this structure has gained widespread acceptance in CADD systems, surveying, engineering, and terrain analysis.

The TIN structure was proposed by Peucker et al. (1976) and takes into account an important property of geographic data collection, that is, that map data collection often tabulates data at points and that the points themselves have an inherent significance as far as the information content is concerned. For example, a surveyor measuring land surface elevations seeks "high information content" points on the landscape, such as mountain peaks, the bottoms of valleys and depressions, and saddle points and break points in slopes, and he or she measures the elevation at these points. The intervening elevations are then often interpolated from the significant points.

The simplest form of interpolation is to assume that between triplets of points the land surface forms a plane. This is the foundation of the TIN structure. Triplets of points forming irregular triangles are connected to form a network. Rules are followed in partitioning the points into the nodes of the triangles. The space is then divided into a set of triangles, their edges, and their nodes. Data can be associated with any of these elements. Two approaches are in common use to arrive at the "best" triangulation. In the first method, an initial triangulation is assigned and is then refined until the triangles meet the Delaunay criterion. Alternatively, some methods seek to compute the best triangulation in one step. McKenna (1987) reviewed these approaches.

The first problem in TIN creation is how to allocate the triangles between points. This problem is usually solved using the Delaunay triangulation (Figure 8.8). This is an iterative solution that begins by searching for the closest two nodes and then assigns additional nodes to the network if the triangles they create satisfy a criterion, such as selecting the next triangle that is closest to a regular equilateral triangle. Many different methods for performing this space partitioning exist and are constantly being improved and made more efficient (McKenna, 1987)

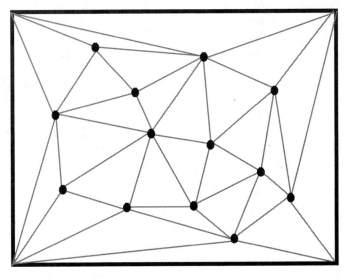

Figure 8. 8 Delaunay triangulation to yield a TIN.

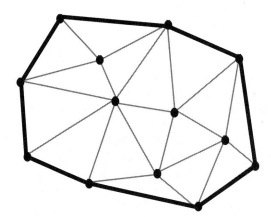

Figure 8.9 A convex hull.

The Delaunay triangulation yields a convex hull (Figure 8.9). This is because if we make an attachment that makes a concavity in the edge of the network, the network can be extended until the concavity is removed. So in other words, there are no concavities in the hull's edges. The convex hull presents another problem in constructing a TIN, because no data can be stored in the structure for the region between the edge of the convex hull and the square or rectangular edges of the map. A common solution is to include the map corners, and perhaps points along the map edges, in the Delaunay triangulation or to go beyond the map edges with additional points outside the area of interest.

Two data structures have been used to physically represent TINs within the computer's memory. The first uses the triangle as the basic cartographic object, while the other uses the vertices of the triangles. Most TIN structures use the triangles as the object, and store links to neighboring triangles (for example, see Gold et al., 1977). In this case, data are contained in two files (Figure 8.10).

The first file contains points, stored as [x,y,z] triplets with their eastings, northings, and elevations. The second file contains one record per triangle and contains three attributes that are pointers into the point file, plus three additional pointers to adjacent triangles. Function 8.1 gives an example of data stored in this data structure.

The original TIN structure used in Poiker and Chrisman's paper (Peucker and Chrisman, 1975), used a vertex-based system. In this data structure, the point record contains [x,y,z] as before, but also a pointer to a "connected points" file. The connected points file is sequential and contains, at each location pointed to in the points file, a list of nodes that are connected to the point in question. The list is terminated with a zero to allow an arbitrary number of nodes (Figure 8.11). This data structure uses about half the storage of the triangle object data structure. Each of these two structures has advantages and disadvantages depending on the particular application. In both cases the second file is constructed from the points file during the Delaunay triangulation phase of the TIN construction.

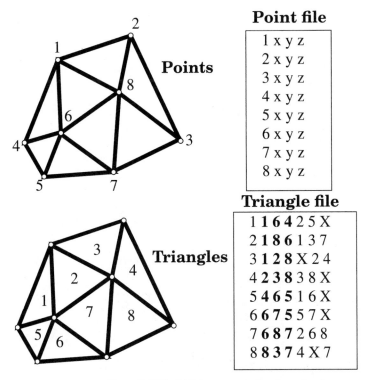

Point file

1 x y z
2 x y z
3 x y z
4 x y z
5 x y z
6 x y z
7 x y z
8 x y z

Triangle file

1	**1 6 4**	2 5 X
2	**1 8 6**	1 3 7
3	**1 2 8**	X 2 4
4	**2 3 8**	3 8 X
5	**4 6 5**	1 6 X
6	**6 7 5**	5 7 X
7	**6 8 7**	2 6 8
8	**8 3 7**	4 X 7

Figure 8.10 The TIN data structure based on triangles.

Function 8.1

```
/* C language structure to hold a TIN */
#define MAXTRIANGLES 20 /* Max # triangles in TIN */
typedef struct {
      POINT Point;
      int elevation;
} TRIPLET;
typedef struct {
      int Vertex[3];
      int Neighbor[3];
} TIN;
TIN Tin[MAXTRIANGLES];
/* the tin array holds the triangle file */
TRIPLET Triplet[MAXPOINTS];
/* the triplet array holds the points file */
```

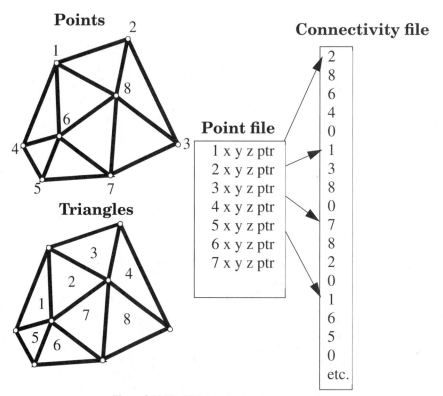

Figure 8.11 The TIN data structure based on vertices.

A major advantage of the TIN structure is that many of the problems associated with automated contouring, hidden-line processing, and surface shading for cartographic symbolization are simpler to tackle in this structure. The structure is particularly suitable for workstations that support polygon manipulation and shading in three dimensions in hardware, allowing interactive production of perspective views and realistic perspectives. The structure is also advantageous for modeling, especially of hill slopes and streams, which flow along the edges of the triangles across the terrain. The TIN data structure is also very suitable for the intervisibility problem, which seeks to determine in three dimensions solutions to which facets in the TIN are visible from points and lines, perhaps at different altitudes.

8.6 QUAD TREE DATA STRUCTURES

The quad tree is a data structure that has received increasing use in recent years. At first, quad trees were used exclusively in image processing for binary images and for partitioning lines. More recently, a number of advances in algorithms have made quad trees a viable data structure for cartographic data. Quad tree equivalent algorithms now exist for area computation, centroid calculation, image comparison, connected component labeling, neighbor detection, distance transformations, regionalization, smoothing, and edge enhancement (Tobler and Chen, 1986). Mark and Lauzon (1985), among others, pointed out the feasibility and advantages of the data structure for map data, and Samet (1984) provided the definitive survey.

Quad trees are areal indexing systems and assume that the basic data element is the area, as represented by the extent of a grid cell. Quad trees, used as a cartographic data structure, also assume that all areas to be stored are grid cells with a side that is some power of 2 of the smallest resolution element. The quad tree data structure allows very rapid area searches and relatively fast display. The gain in efficiency is the result of the exploitation of the fact that areas on maps are typically some combination of large homogeneous areas and smaller heterogeneous areas. Quad trees are tessellation data structures, in that the space on the map is partitioned by a covering network of spaces, but for the quad tree, the partitioning is into nested squares. Division of squares is into quadrants, giving the "quad" part of the name.

At the highest level, the entire map can be thought of as one square. Rectangular or irregular maps can be subdivided into a set of square regions to become starting points in the quad tree. Each square is then divided into four quadrants, the NE, NW, SE, and SW, if and only if the square contains detail at a level greater than the current area of the square. The path to each quadrant then becomes a branch, from the root of the quad tree to its leaves, which are the highest level of detail.

This system is familiar to users of the U.S. Public Land Survey System. In this system, square regions (and rectangular, and indeed irregular shapes due to errors) are given a reference within a township and range system. For example, a small land tract may have the following reference

```
Northeast Quarter of the Southwest Quarter of the Southeast
     Quarter of Section 21, Township 24N, Range 2E.
```

When the land holding is large, it can be referenced simply; for example, we could record that all of Section 21, Township 24N, Range 2E is farmland. When the landholdings are small and various, we need the full list to specify the plot.

To index a very small area within a quad tree structure, we could use a list such as NE, SW, NE, NW, SE, and so forth.

More efficient, however, is to store the Morton number. The Morton number is a unique identifier assigned to quad tree divisions as they are constructed. Figure 8.12 shows how Morton numbers are assigned. In their simplest form, quad trees consist of strings of Morton numbers, each of which contains the index or pointer to a data value associated with the highest-level quadrant. When large squares are sufficient, the Morton sequence is short, and vice versa.

Quad trees are most efficient when the basic cartographic entity is the polygon. The cells in the quad tree then become the cartographic objects, and the data structures contain the coordinates of the region in question, the Morton sequences for each unique cell division in the quad tree, and the attributes for these cells. More than this, the efficiency of the quad tree is at its highest when a mix of large homogeneous areas and smaller polygons exists, because the storage gain is mostly from two-dimensional run-length encoding of the large regions.

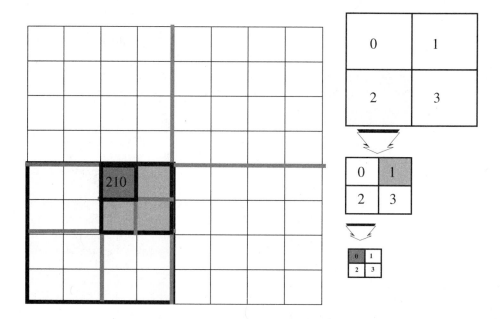

Figure 8.12 Selection of Morton code 210.

Land-use maps are good examples, when this is the case, as also are political units such as counties. Land-use maps generally have a background category that occupies most of the space, such as forest or agriculture. The map then has patches on top of this background, such as roads, commercial districts, lakes, and wetlands. In these cases the quad tree may be the most effective structure for several types of analysis to be performed on the data. Given that many applications in remote sensing generate regionalized or "segmented" grid maps, the strong link among the quad tree, remote sensing, and the grid data structure is self-evident.

8.7 MAPS AS MATRICES

The logical organization of raster data is easily converted into a physical representation in storage. This is because almost all computer programming languages support the array as a data structure directly. Thus in C, the array is given as z[i][j], in Pascal as z[i,j], and in FORTRAN as Z(I,J). As such, the mathematical term for this organization is *matrix*. In simple terms, we can consider a map to be a matrix. Each element in the matrix falls as a grid cell on the ground and can be set to coincide with the spacing and orientation of the coordinate system and map projection in use. For example, we could divide the world up into grid cells based on 10-degree increments of latitude and longitude and store data for the cells in a matrix. Given the corners of the matrix, and the

```
grid.z[i][j]
where (0 <= i < nrows) and (0 <= j < ncols)
```

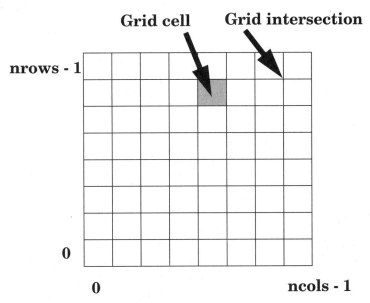

Figure 8.13 Generic structure for a grid.

spacing, we can implicitly refer to each matrix element just by referring to its row and column numbers, rather than the explicit way we give eastings and northings for points (Figure 8.13).

When we read a matrix into the computer's memory, we read it row by row and column by column, one element at a time. Having the grid stored in this way has the distinct advantage of simplicity. The disadvantage is that elements that are spatially next to each other are not necessarily close together in the file. Fortunately, when the data are within the array in a program, the immediate neighbors can be found easily.

Most automated mapping systems that use pixel-based display or use remotely sensed data in any form use the grid as their basic data structure. Usually, the files actually stored contain the registration information, such as the coordinates of the corners, the grid spacing, and so forth, plus the image itself, often in binary or compressed form to save storage. Files can be very large using the grid structure. For example, many display devices support images of 512 rows by 512 columns. A single data set, storing one byte per record, takes up 262 kilobytes without compression.

Both the Spatial Data Transfer Standard and GKS define a basic grid as a primitive element. In SDTS, the elements are the grid cell and the pixel, and the composite object of an image. Under GKS, the elements are the cell array and the pixel. As in the data standards, the GKS cell array grid cell need not be a single pixel. Using these elements in

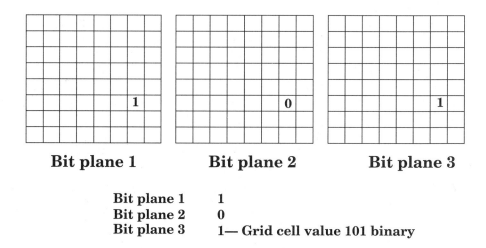

Bit plane 1   Bit plane 2   Bit plane 3

Bit plane 1	1
Bit plane 2	0
Bit plane 3	1— Grid cell value 101 binary

Figure 8.14 Bit plane storage of a grid map.

GKS is termed *raster graphics*, defined as a display image composed of an array of pixels arranged in rows and columns. Because this terminology is part of the standard, the issue of raster or vector representation is trivial and either type can be displayed on any device

A common format for the provision of raw grid data is the DEM format of the U. S. Geological Survey. A program to read the DEM format, as the files are available on tape or over the Internet, is provided on the companion disk as `read_dem`, and a sample data set is provided. The program `draw_image` reads the grid files output by this program, converts the grid to an image, and uses GKS to plot the maps. It is covered in detail in Chapter 12. Examples of data stored in a grid format have been given in Chapter 7, and include the DEM. Issues associated with grid data are registration problems, map projections, boundary effects, and edge matching, especially when the map edges do not neatly fit along the edges of the grid. The problem of piecing multiple grid files together is called mosaicking.

An alternative way of storing information in an array is to use *bit planes* (Figure 8.14). Bit planes are much more associated with storage and display than with manipulation, and are particularly useful for large areas with similar values, although they are universally applicable. In some cases, bit-mapped terminals can be used directly using this structure. Very high resolution monochrome terminals are often bit-mapped.

Bit planes can be used in two ways. Bit planes can represent colors on a display as a map of the pixels. One bit plane can be made to correspond to each color. For the six major colors (red, green, blue, magenta, cyan, and yellow), we need only three bit planes. For a single cell, if all three planes are 0, the screen shows black. If all three planes are 1, the screen shows white. Red, green, and blue correspond to the three planes directly, with a 1 in the correct plane determining the color. Magenta has a 1 in red and blue bit planes, cyan has a 1 in the green and blue bit planes, and yellow has a 1 in green and red.

With three bit planes, we can store eight colors, and because the bit arrays can be placed directly into the video display channels, a color screen can be drawn quickly. Four and more bit planes become complex to use, although storing data bit by bit in bit planes allows extremely fast overlay of images and permits binary operations between planes, such as AND, OR, and NOR. Each bit plane, in addition, can be run-length encoded to drastically reduce the amount of memory required (Figure 8.15).In run-length encoding, we store pairs of bytes, the first representing the value stored in the array, the second giving the number of times that the value should be repeated. Major space savings are possible using run-length encoding when the map is spatially homogeneous.

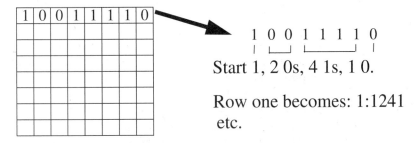

Start 1, 2 0s, 4 1s, 1 0.

Row one becomes: 1:1241 etc.

Figure 8.15 Run-length encoding for a bit map.

8.8 MAP AND ATTRIBUTE DATA STRUCTURES

In this chapter, we surveyed the major data structures in use for the storage of cartographic objects within their mapping context. Analytical and computer cartography, however, requires far more than simply the map data to produce an effective map, to use the data for new cartographic applications, and to support the demands of GISs. In many ways, map data structures are the determining factor for computer cartography. Many of the issues of geocoding, data conversion, error control, analytical power, and the choice of map display are really determined by the map data structure used. Understanding the strengths and weaknesses of these structures is critical to the understanding of analytical cartography.

In Chapter 9 we will consider how attribute data can be similarly structured and examine the need for effective links between the map and the attribute data. It is on the strength of the map to attribute link that many mapping systems and GISs will stand or fall as new and innovative uses for cartographic data find their way into the everyday life of noncartographers. We should remember, therefore, that a good cartographic data structure must both preserve the data integrity and also make the geographic information it contains available to make maps that can be used and understood by all.

8.9 REFERENCES

Anstaett, M. R., and H. Moellering (1985). "Area Calculations Using Pick's Theorem on Freeman-encoded Polygons in Cartographic Systems." *Proceedings, AUTO-CARTO 7,* Seventh International Symposium on Computer-Assisted Cartography, Washington, DC, March 11–14, pp. 11–21.

Burrough, P. A. (1986). *Principles of Geographical Information Systems for Land Resources Assessment.* Monographs on Soil and Resources Survey Number 12, Oxford Science Publications. Oxford: Clarendon Press.

Dutton, G., ed. (1979). *Harvard Papers on Geographic Information Systems.* First International Advanced Study Symposium on Topological Data Structures for Geographic Information Systems. Reading, MA: Addison-Wesley.

Gold, C. M., T. D. Charters, and J. Ramsden (1977). "Automated Contour Mapping Using Triangular Element Data Structures and an Interpolant Over Each Irregular Triangular Domain." *Computer Graphics,* vol. 11, pp. 170–175.

Mark D. M. (1975). "Computer Analysis of Topography: A Comparison of Terrain Storage Methods." *Geografiska Annaler,* vol. 57, pp. 179–188.

Mark, D. M., and J. P. Lauzon (1985). "Approaches for Quadtree-based Geographic Information Systems at Continental or Global Scales." *Proceedings, AUTOCARTO 7,* Seventh International Symposium on Computer-Assisted Cartography, Washington, DC, March 11–14, pp. 355–364.

McCullagh, M. J., and C. G. Ross (1980). "Delaunay Triangulation of a Random Data Set for Isarithmic Mapping." *Cartographic Journal,* vol. 17, no. 2, pp. 93–99.

McKenna, D. G. (1987). "The Inward Spiral Method: An Improved TIN Generation Technique and Data Structure for Land Planning Applications." *Proceedings, AUTOCARTO 8,* Eighth International Symposium on Computer-Assisted Cartography, Baltimore, MD, March 29–April 3, pp. 670–679.

Peucker, T. K., and N. Chrisman (1975). "Cartographic Data Structures. "*American Cartographer,* vol. 2, no. 1, pp. 55–69.

Peucker, T. K., R. J. Fowler, J. J. Little, and D. M. Mark (1976). *Digital Representation of Three-dimensional Surfaces by Triangulated Irregular Networks (TIN).* Technical Report No. 10, U.S. Office of Naval Research, Geography Programs.

Peuquet, D. J. (1979). "Raster Processing: An Alternative Approach to Automated Cartographic Data Handling." *American Cartographer,* vol. 6, pp. 129–139.

Samet, H. (1984). "The Quadtree and Related Hierarchical Data Structures." *IEEE Transactions on Pattern Analysis and Machine Intelligence,* vol. PAMI-4, no. 3, pp. 298–303.

Tobler, W. R., and Z. T. Chen (1986). "A Quadtree for Global Information Storage." *Geographical Analysis,* vol. 18, no. 4, pp. 360–371.

PART III
Analytical Cartography

Analytical cartography consists of the basic mathematical and algorithmic principles of cartography that survive independently of the particular technology used to create maps. This part contains five chapters that introduce the transformational approach to cartography and then give examples of the various types of transformations. Chapter 9 covers data models for the attribute or nonmap data associated with management of GIS-type data for computer mapping. Chapter 10 provides the background discussion for the section, introducing the ideas of Tobler and Unwin on map types and map transformation classification frameworks. Chapter 11 covers transformations of the map data, including generalization and geometric transformations of the graphic depiction. Chapter 12 covers the implications of the transformations for moving between computer object representations of cartographic data, that is, between data structures. Chapter 13 looks specifically at the problems of representing and analyzing terrain using the transformational paradigm. Throughout, examples of the transformations are developed as computer programs, consistent with the data structures introduced in Part II.

9

Attribute Data Structures

9.1 SEQUENCE OF DEVELOPMENT

In the preceding chapters we examined physical data structures, the actual arrangement of digital cartographic objects in computer files. We also examined logical data structures, the conceptual links between digital cartographic objects and their physical data structures. Since the advent of geographic information systems (GISs), much attention has been devoted to how effective the physical-to-logical data structure link is, especially when using these structures for interactive query and analysis. These additional functions (additional, that is, to computer cartographic systems) require the ability to query thematic data using geographic properties. More strictly we query the cartographic objects representing the geographic phenomena by their attributes reflecting geographic properties. In addition, these queries require the retrieval of data from many different storage locations, usually many files with different sizes and formats.

Advanced analytical operations, the sort supported in GISs, require the management of attribute data to support query and analysis by cartographic data structures. The map and the attribute data physically reside in separate files or are separated within the computer's memory. The problem of managing these files is one of database management, a field well represented within computer science. If we look at the sequence of how database management systems developed, we find that there are distinctive trends in the historic development of attribute data structures. Although these trends are more important to GISs than to computer cartography, they are of relevance to analytical cartography and therefore will be discussed here.

The first generation of GISs and computer cartographic systems did not really support query and analysis interactively. These early systems had sophisticated map data structures, but were usually grid or entity-by-entity based, and often included the thematic data as part of the cartographic data structure. Because the purpose of these systems was to produce a map, the independence of the data was sacrificed. For example, to convert choropleth data from numbers to percentages would usually involve reentering the thematic data, perhaps as a set of parameters on "cards" for batch processing. The multifile problem was avoided by merging the data into a single file containing the data structure.

Second-generation systems were still entity-by-entity structured, but they were topological systems. Using these systems, we could expand the basic entity of a point to build lines and then lines to build polygons. Polygons in a topological system could consist of a chain list or a point list, but we also have as part of the data structure information about which chains are linked to which other chains. By keeping track of forward and reverse linkages so that we can reassemble our polygons using the topology, if we need to fill them or verify them, we are actually storing attribute data as raw geographic properties, in particular connectivity and contiguity. If we need to know which lines are connected to other lines, we can find that out from our data structure.

In these systems, different parts of the structure could be stored in different files. As a result, the data structure is more flexible. The thematic data, however, often remained as part of the map information. For example, the DIME files consisted of single sets of records, with index numbers for the polygons. The user of the DIME files had to manually merge the census variables with the map data to produce computer-generated maps. The only physical link between the topological and the attribute data was the census tract or block number, which happened to be the same in each of two independent data sets. Although this arrangement was closer to supporting query and analysis, the data structure itself was not the means by which query was performed.

Third-generation systems combine map and attribute data together, usually within a GIS. Third-generation systems are distinctive because they are capable of performing separate data management, that is, performing transformations on the thematic data directly without reference to the geography and across multiple files. These systems have been able to integrate the capabilities of database management systems, such as selective retrieval, fast sorting, and operations on multiple data attributes, into systems designed for the display and analysis of geographic data. These systems use the concept of the data model and implement one of three basic models common in database theory.

The first two of these data models, the hierarchical and network models, have now been largely superceded, but are important historically and so are considered in section 9.2. The third of these models, the relational model, is itself a movement to a fourth stage of attribute data structure development. As we will see, it is this data model that has allowed the development of GISs capable of performing real-time query and analysis by geographic properties and their pertinent cartographic objects. It is, therefore, worth examining the basic data models before returning to an examination of some current attribute data structures found in contemporary GISs.

9.2 ATTRIBUTE DATA MODELS

A data model can be defined at any one of three levels. The data to be modeled correspond to the geographic phenomena or entities as they exist in the real world. Because they are data, they correspond to a set of measurements of real-world phenomena. The three levels at which data models have been designed are (1) the conceptual level, at which level the model is independent of the specific systems or program data structures used to organize or manage the data; (2) the logical level, which corresponds to the data structure, or logical organization of the components of the model and in which the relationship between components are explicitly defined; and (3) the physical level, at which a set of rules exist which specify how the machine implements the data structure within a computing environment (Peuquet, 1984).

This division corresponds closely to the terms entity, object, and data structure used in this text, although the data structure has been shown to contain both a logical (design) and a physical (file or programming language) structure of its own. The latter division is largely artificial, to understand the data structures in use for spatial data it is necessary to implement them, or at least understand their implementations. For attribute models, however, it is less critical to understand the physical implementation. The discussion here, therefore, is restricted to the conceptual data model level in Peuquet's hierarchy.

9.2.1 The Hierarchical Data Model

As discussed previously, in the management of attribute data, the equivalent of the term *cartographic data structure* is the term *data model*. Data models were developed from database management theory, part of computer science. The first and simplest of the data models is the hierarchical model. The hierarchical way to structure data can refer to many different attributes, not just those of geographic data. Hierarchical data management systems form the basis of many commercially available systems and have influenced file structures and our thinking about organizing data generally.

To use a geographical example, though, the top of a hierarchy is called the *root* (from the root of a tree) and for geography the root could be the WORLD (Figure 9.1). We know that there are seven CONTINENTS, and one of those continents is North America, which consists of several COUNTRIES, one of which is the United States of America. The United States consists of 50 STATES, one of which is New York, and New York consists of COUNTIES, one of which is New York (same name, but different level in the hierarchy). New York County, otherwise known as Manhattan, contains DISTRICTS, one of which is the Upper East Side, where we can find the LOCATION of Hunter College.

This is a hierarchical structure. Each piece of the structure has links to "children" and a single "parent" through which all searching must pass. A full reference to a particular element, such as a cartographic entity like Hunter College, a unit upon which we are going to hang data, gives its path down the "tree." An atlas is usually structured this way, with all maps for Asia bound together and separated country by country. At the lowest level of the tree, if we think of the tree as inverted, we find the "leaves" or records. For cartographic data the records are most commonly cartographic objects, and we assume that the map database contains graphic primitives associated with the cartographic objects. In addition, normally for each entity we have attribute information.

Attributes can store nominal, ordinal, interval, or ratio data, as well as text or complex characteristics. Attributes and entities have both a physical and a logical structure. In database management systems, the physical structure is often hidden. Logically, the attributes and entities can be thought of as occupying a flat file.

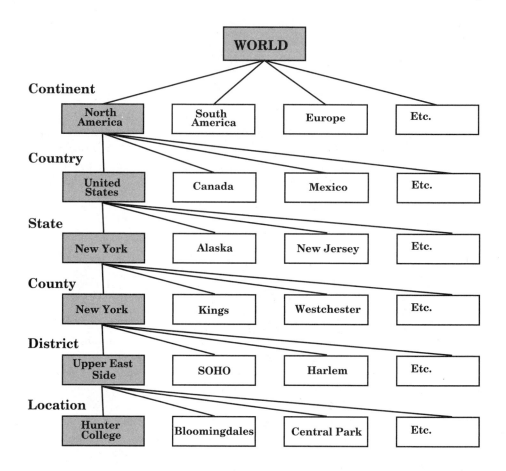

Figure 9.1 A geographic hierarchy.

A flat file (Figure 9.2) resembles a spread-sheet table or matrix, with rows of records and columns of attributes. A flat file corresponds to a particular level of the hierarchy. For example, a flat file for New York State could contain counties as records, with each county having another flat file further down the hierarchy. The census distribution tapes and the city and county data book are good examples of flat files. Usually, all the communities are listed as rows, and hundreds and hundreds of data values for different attributes are listed as columns.

In the hierarchical data model, we organize the links between the logical and the physical aspects of the data in a tree structure. Although this system is often adequate, several problems can result. A major problem arises from cartographic objects that belong to multiple hierarchies. For example, police districts, school districts, census tracts, voting districts, and congressional districts rarely coincide, all being part of different hierarchies. Yet a single cartographic entity falls into all these areas. Also, hierarchies are

Attributes

	Mountain	Elevation	Country	Second Country	Etc
E n t i t i e s	Everest	8847	Nepal	China	
	K2	8611	China	Pakistan	
	Annapurna	8078	Nepal	——	
	Dhaulagiri	8172	Nepal	——	
	Gasherbrum	8068	China	Pakistan	
	Kanchenjunga	8598	India	Nepal	
	Makalu	8481	China	Nepal	
	Nanga Parbat	8123	Pakistan	——	
	Xixabangma Feng	8012	China	——	

Figure 9.2 A flat file.

rarely spatially mutually exclusive, leaving gaps in some areas and double covering other areas. For example, the boundary between New York and New Jersey is marked at the halfway point on the George Washington Bridge. If the bridge is a single cartographic object in a cartographic database, which state is it included under? Also, while cities are usually parts of counties, New York City contains five counties, an inversion of the usual direction of the hierarchy.

A simple geographic type of query of a hierarchically structured attribute data base might be: How many people live in Manhattan community district number 4? Does it cross any census tract boundaries? An example of hierarchical cartographic information is contour lines on a map. The 100-meter contour line presumably contains all points with elevations greater than 100 meters, unless annotated as a closed depression. Hierarchical space partitioning may therefore be suitable for map data but unsuitable for all but the most rigidly encoded attribute data. The method also involves duplication and therefore is not highly storage efficient.

9.2.2 The Network Data Model

The second data model that has been used in database management theory is the network model. In the network model, we allow relationships between entities. Normally, these are not complex relationships. For example, we might allow an entity, a particular block in Manhattan, to belong to a certain congressional district, a census tract, and a school district. Here, the entity is a block and the relationship is a link to an area.

In computer science, a particularly powerful data structure is the linked list. In a linked list, each member of a list contains a pointer to the next member of the list. We saw in Chapter 5, for example, in the USGS's GIRAS files, that the polygon files simply contain sequential pointers to records in the arc files, which in turn contain pointers to the coordinate files. This is the linked list concept, and this concept is behind the network structure.

Links between objects in this structure form networks. Geographically, the pointers carry associations such as ". . . is a member of." Using the parent-child analog above, network systems allow children to have multiple parents, and even allow a child to be its own grandparent. Thus we can consider the hierarchical to be a subset of the network model.

We remove much of the structure of the hierarchy altogether and let our most common, or the highest resolution entity, become the basis for our model. Although this is true geographically, in a attribute context it does not matter if we do not know the exact extent of a particular area, as long as we store pointers from that entity to everything that incorporates it. This model has won a fair degree of acceptance as a data structure for geographic information, although the query ability of these systems may be as limited as that of hierarchical systems (Aronson, 1987).

9.2.3 The Relational Data Model

Returning to the context mentioned above, of levels of development of data management within cartographic and geographic analytical systems, we have now reached the fourth-generation system. In some fourth-generation systems, the data model is termed the *relational model*. The basic unit within a relational system is the table. Within each table we store the name of an entity, the type of that entity, and the attributes associated with that entity. Each table closely resembles the flat file mentioned previously.

One of the attributes associated with a cartographic object, a line, for example, would be the points it contains. Within a relational structure, the points would be stored sequentially in a separate file, because normally relational systems do not use ordering of records. The link between the files is by a *key* or index. The index could be a line identifier. So, for example, in a DLG file we could have a particular line representing a part of an interstate highway between two junctions. On the map, the line would be symbolized as a red double line with a black border, a fact that could be stored along with the line attributes.

The attributes of the line, however, need not be stored with the line. The line could be numbered with a key, and the corresponding key link to another file could contain the attributes of the line, for example, that it is an interstate highway, the type of road surface, the volume of traffic, and so forth. An additional attribute may also be a further index, for example, links (keys) to a file of areas, such as counties, through which the road passes. At this level, links can be established freely between any of the objects, and only the minimum number of files is required to store the data.

A graphic data structure with relational characteristics is very versatile. We can come into this file structure anywhere, at any point, and retrieve useful information. For the interstate highway, we may have a single arc record and wish to make a query at the arc level. One thing we might be interested in is what arcs meet this particular arc at the end of this segment. A route-finding application, for example, may need to provide in-vehicle instructions based on the connecting highways. All we need do is follow the segment along to the end, sort the arc attribute file, and from that retrieve the polygon identifiers that contain these arcs, and so find the polygons that include a particular location. Thus the in-vehicle system could notify the driver as she or he passes from town to town, city to city, or county to county.

The relational structure involves some duplication of information between data files. Also, we end up doing a tremendous amount of indexing. The common element within relational structures is the key, sometimes called the index, which is a unique identifier shared between two files. In relational database management systems complete software systems take care of retrieval of records based on their attributes and even contain systems called data definition languages for setting up new tables and entering data into the tables. Many such systems are now commercially available, and several have been used effectively for geographic data. Relational database management systems also minimize data volume by a process called weeding. As we build new files, and it is very easy to build files out of existing files in relational systems, the system automatically looks for duplication of keys within a file and eliminates redundant records.

Query within a relational system allows some very powerful and flexible searches such as the joining of data sets to provide temporary sets useful for a particular query. Aronson (1987) noted that the relational structure is most amenable to geographic data because of its simplicity, flexibility, efficiency of storage, and its nonprocedural nature, that is, its free-form query. The relational data model emerged as "the dominant commercial data management tool of the eighties" (Aronson, 1987).

9.3 SOME ATTRIBUTE DATA MODELS AND STRUCTURES

9.3.1 The Hypergraph Structure

The way the relational data model has found its way into GISs is generally in its extended form as the *entity-relationship model,* which is now the basis of several successful GIS systems, such as the ARC/INFO system (Morehouse, 1985). To some degree, the entity-relationship concept is similar to an idea by Francois Bouille, a French computer scientist who designed the hypergraph approach and successfully introduced the idea into the English language literature on GIS. The approach suggested that a body of theory from computer science and relational database management theory could be linked by two mathematical subfields, set theory and topology, then applied to geographical data.

Bouille proposed in 1978 the four fundamental geographic concepts of objects, class, attributes, and relations (Bouille, 1978). We can think of these as cartographic objects, dimension, attributes, and linkages. In addition, Bouille used abstract data types: object class, attribute of object, attribute of class, relationships between objects, and relationships between classes. Under this system, it was possible to develop an *n*-dimensional diagram that explicitly stated all the possible links between the objects. Its expression in topological space was a concept borrowed from the mathematical field of topology, called a hypergraph.

A *hypergraph* is a network consisting of nodes and edges. The hypergraph provided the mathematical basis for making relations link objects, classes, and their attributes. The entity-relationship model uses similar ideas, as seen below. There are examples of implementations of the hypergraph structure, to road networks, for example (Rugg, 1983). The idea led to considerable activity in publishing and research in the area of relational database modeling for spatial data.

9.3.2 The Entity-Relationship Structure

Nyerges introduced into GIS and cartography a model developed by Chen (1976), which Chen termed the *entity-relationship model* (Nyerges, 1980). A very similar model was proposed in the same year by the computer scientists Shapiro and Haralick (1980) and placed into actual use by 1982. This represents a very short gap between research and implementation of new systems, a sign of a vigorous research frontier.

Chen noted that his model was more powerful than the simple relational model and that it had all the capabilities of both the network and the relational models. The entity-relationship model has become the standard for many recently developed GISs. In many respects, this is the "single super-flexible structure forming the basis of geographic data management, display and analysis" (Clarke, 1986). An extension of the entity-relationship design is the relational model with non–fixed-length sets of attributes. In computer science, such data management has come to be called object-oriented, and a number of systems use the approach. So far, only a few GISs use this method, although the concept is pertinent to many previous designs.

9.3.3 Object-Oriented Models

A considerable amount of interest has been devoted recently to the application of the tools of object-oriented computer programming (OOP) to data bases. Object-oriented programs use extensions of the programming language to support *classes*. A class is somewhat like a structure, that is, it can contain multiple pieces of information about a single object. Data elements that fit into classes are called *objects,* hence the name object-oriented. A class specification within OOP represents a sort of template, a generic set of attributes and features that an object can possess.

More sophisticated capabilities are supported by OOP systems, however. A class of objects can have a set of attributes, measures, algorithms, or operations that apply only if the object meets its formal specifications. For example, to compute an intersection point between two line segments, we must have two line objects that are not parallel or coincident, and that meet at a point. If these conditions are met, the "rules" or the algorithm for line intersection can be thought of as "applying" to that object. Passing a template from one object to another, therefore, passes on all the rules also. This is called *inheritance* (Figure 9.3).

Thus complex objects can be made from simpler objects by inheriting their specifications, or new instances of objects can be created by copying the template from another object. The particular set of measures, conditions, and rules that apply to an object can be stored, manipulated, and even passed along with the object to other parts of the program,

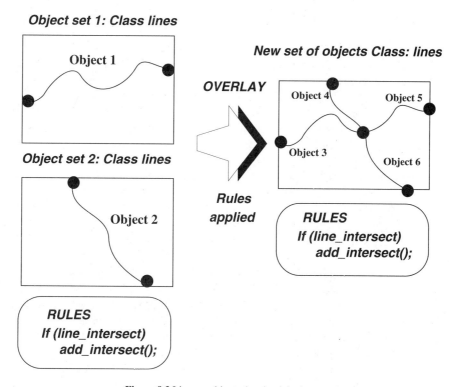

Figure 9.3 Lines as objects showing inheritance.

a condition known as *encapsulation*. The user can control how much of the encapsulated data need be bundled with the object and never revealed and how much needs to move along when the object is modified or used elsewhere in the data base.

A particular object, placed into the template is termed an *instance* of that object. In Figure 9.3, two instances of an object of the class line exist and are encapsulated with their set of intersection rules. An overlay operation then creates a new set of instances of the class line, generated by breaking the objects according to the rules. Each of the new objects inherits the class line and the rules associated with it.

Once a full set of generic objects is defined, each object can simply be treated as one data entry in a very simple data base. The use of object-oriented databases, therefore, allows a whole range of checking, automatic recomputation, reordering and so forth that can be *hidden* from the user. The user of a database, for example, need not know that a set of lines and polygons stored in the data base "know" how to insert new segment points when intersection occurs in the overlay process. The user can simply perform the database operation to do the overlay, without being concerned about each step along the way.

The costs of using object-oriented systems are twofold. First, before beginning, the user must spend time building a full set of objects and relations between objects for use with the data. This is very difficult, and few cartographers to date have attempted this other than in specific contexts, despite that spatial objects and their relations are apparently well defined and understood. In addition, the full overhead of a full set of operations and relations is attached to every data capsule, generating the need for powerful systems to implement the methods. On the other hand, once these object specifications are built, they can be reused again and again for different projects, data, and so forth.

A new generation of software is now appearing that uses object-oriented programming and object-oriented databases for cartographic purposes. A distinct advantage is the full encapsulation of all the data and rules about objects such as map features, giving considerable analytical flexibility to these systems. The systems need no longer make distinctions between graphics, attributes, and algorithms as far as cartographic data are concerned. The impact of this fact, that software and algorithms can become as generic as the data structures used for map data, will be quite liberating for analytical and computer cartography. The powerful extensions to the C programming language in C++ have become the major tools with which these systems are being constructed.

9.3.4 Hybrid Structures

Roger Tomlinson in 1978 argued that geographic data will never conform to the computer science data models and that we should not try to force geographic data into structures that were designed for handling attribute data. Tomlinson noted that in mapping and analysis we have two different goals: In mapping, all we really need to do is retrieve the entities and the geographic locational attributes associated with those entities. In analysis we very often want to retrieve both the entities and the relationships between the entities. Tomlinson said that it might be more appropriate to develop entirely new data models from a geographic viewpoint. A "geographic" data model has yet to be developed.

As far as cartography is concerned, such a model may never need to exist. Only in GISs do we need to embed more information than we need to depict graphically a map as a set of symbols, and to support cartographic analysis. A "geographic data model" may not be the answer as far as GIS is concerned because no matter how complex the data structure, there are always queries that reveal weakness in the data structure for a particular application.

An alternative is to look at some other approaches that have less flexibility. One approach has been to propose systems with "artificial intelligence," or at least with "expert knowledge," which interact in a more effective manner with the GIS user and structure the data accordingly (for a survey, see Robinson and Frank, 1987).

Finally, we could use the existing approach, which seems to be to build a new or modified data structure for each new application. Although the off-the-shelf nature of database management systems has made this increasingly easy over the last few years, it is the nature of the map to attribute data link that is of most importance. Systems that support real geographic query allow the user to interact with the data through the map rather than though a menu or command-line type of interface. A cartographic "window" into the data is essential, as is the ability to view and manipulate cartographic symbols while managing the data.

Clearly, attribute data structures are essential to GISs. On a continuum, they are also vital to analytical cartography, where the management of attributes is critical, but less so for computer cartography, where it is sufficient simply to retrieve the correct attribute. It is appropriate, therefore, to conclude this part on the representation of cartographic data by examining the utility of the attribute data models for analytical and computer cartography.

9.4 DATA STRUCTURES AND MODELS FOR ANALYTICAL AND COMPUTER CARTOGRAPHY

Much of the previous discussion has been appropriate for GISs. What about analytical and computer cartography? Automated cartographic systems usually need only to retrieve entities and display them. The most important attributes are locational, and the need for complex statistical operations is minimal. For many map applications, we would be quite happy with a very simple attribute data structure as far as symbolization is concerned. Because more sophisticated cartographic applications such as dynamic maps and vehicle navigation systems require some kind of attribute structure, the two have to be irreversibly linked. Much of the structure of a map is almost totally independent of the locational attributes that a map portrays.

Analytical cartography, and also computer cartography, should concern itself with the management of attribute as well as map data. Although the map is the primary vehicle for the communication of geographic properties and distributions, it is the effectiveness of the analytical operation, the combination of data structure and cartographic transformation, that increasingly influences which map the map reader has access to. A system that produces the wrong map, or a suboptimal map, for the task at hand may be as poor a piece of cartographic software as that which uses poor design or inappropriate symbols.

The power of both map and attribute data structures, as far as analytical and computer cartography is concerned, is in how readily they support cartographic transformations, a theme discussed in Part III. For many reasons, cartographic transformations, such as particular symbolization methods, require data to be in a specific data structure. In this case, we can add to the power of our structure by choosing those that allow us to transform between structures with minimal error and data loss, or at least error of a known magnitude and distribution.

Chapter 10 introduces the cartographic transformation, and Chapter 11 extends the idea to cover the four major mapping transformations. A particularly important chapter, given the previous discussion, is Chapter 11, in which transformations between structures are considered in more detail.

9.5 REFERENCES

Aronson, P. (1987). "Attribute Handling for Geographic Information Systems." *Proceedings, AUTOCARTO 8,* Eighth International Symposium on Computer-Assisted Cartography, Baltimore, March 29-April 3, pp. 346–355.

Bouille, F. (1978). "Structuring Cartographic Data and Spatial Processes with the Hypergraph-based Data Structure," in *First International Symposium on Topological Data Structures for GIS,* edited by G. Dutton. Cambridge, MA: Harvard University. Laboratory for Computer Graphics and Spatial Analysis.

Chen, P. P.-S. (1976). "The Entity-Relationship Model—Toward a Unified View of Data." *ACM Transactions on Database Systems,* vol. 1, no. 1, pp. 9–36.

Clarke, K. C. (1986). "Recent Trends in Geographic Information System Research." *Geo-Processing,* vol. 3, pp. 1–15.

Morehouse, S. (1985). "ARC/INFO: A Geo-relational Model for Spatial Information", *Proceedings, AUTOCARTO 7,* Seventh International Symposium on Computer-Assisted Cartography, Washington, DC, March 11–14, pp. 388–397.

Nyerges, T. L. (1980). "Representing Spatial Properties in Cartographic Databases", *Proceedings, ACSM Technical Meeting,* St. Louis, pp. 29–41.

Peuquet, D. J. (1984). "A Conceptual Framework and Comparison of Spatial Data Models." *Cartographica,* vol. 21, no. 4, pp. 66–113.

Robinson, V. B., and A. Frank (1987). "Expert Systems Applied to Problems in Geographic Information Systems: Introduction, Review, and Prospects." *Proceedings, AUTOCARTO 8,* Eighth International Symposium on Computer-Assisted Cartography, Baltimore, March 29–April 3, pp. 510–519.

Rugg, R. D. (1983). "Building a Hypergraph-based Data Structure: The Example of Census Geography and the Road System." *Proceedings, AUTOCARTO 6,* Sixth International Symposium on Computer-Assisted Cartography, Ottawa, October 16–21, vol. 2, pp. 211–220.

Shapiro L. G., and R. M. Haralick (1980). "A Spatial Data Structure." *Geo-Processing,* vol. 1, no. 3, pp. 313–338.

Tomlinson, R. F. (1978). "Difficulties Iinherent in Organizing Earth Data in a Storage Form Suitable for Query." *Proceedings, AUTOCARTO 3,* Third International Symposium on Computer-Assisted Cartography, San Francisco, January 16–20, pp. 181–201.

10

A Transformational View
of Cartography

10.1 A FRAMEWORK FOR TRANSFORMATIONS

Much of the focus of the previous chapters has been on the tools of mapmaking and on the structure of cartographic data at each stage in the mapping process. The approach has concentrated upon the *states* of the cartographic data as the data move from cartographic entity to cartographic object to cartographic symbol. But this is only half the story. It is time to ask how we move between states, to focus on the cartographic transformation, and to take *a transformational view of cartography*. In this approach we will concentrate on *analytical* rather than computer cartography. Cartographic transformations are central to what analytical cartography is all about. In fact, analytical cartography could be called a discipline in transformations.

Cartographic transformations come in many forms. There are transformations of attribute data, transformations of the locational properties of maps, graphic transformations, transformations of the information content of maps, and the scale transformations of generalization and selection. As will be shown, it is particularly interesting to ask if any given transformation is invertible, that is, whether or not a cartographic transformation can be undone or reversed to produce the initial starting conditions. Recent advances in mathematical theory have focused attention on transformation inversions that are unstable, that is, the inverse transformation produces chaos. A thorough knowledge of stable versus unstable cartographic transformations would go a long way toward providing a theory of cartographic transformations. Stable transformations are controllable and therefore are effectively programmed and modeled, especially with respect to the error introduced. The analytical cartography introduced in this part is the core of what stands behind much of the computer cartographic software developed to date. This means that it will also be reflected in the computer cartography of the future.

Many of the ideas in this chapter have been adapted from the paper titled "A Transformational View of Cartography," in which the cartographer Waldo Tobler stated his

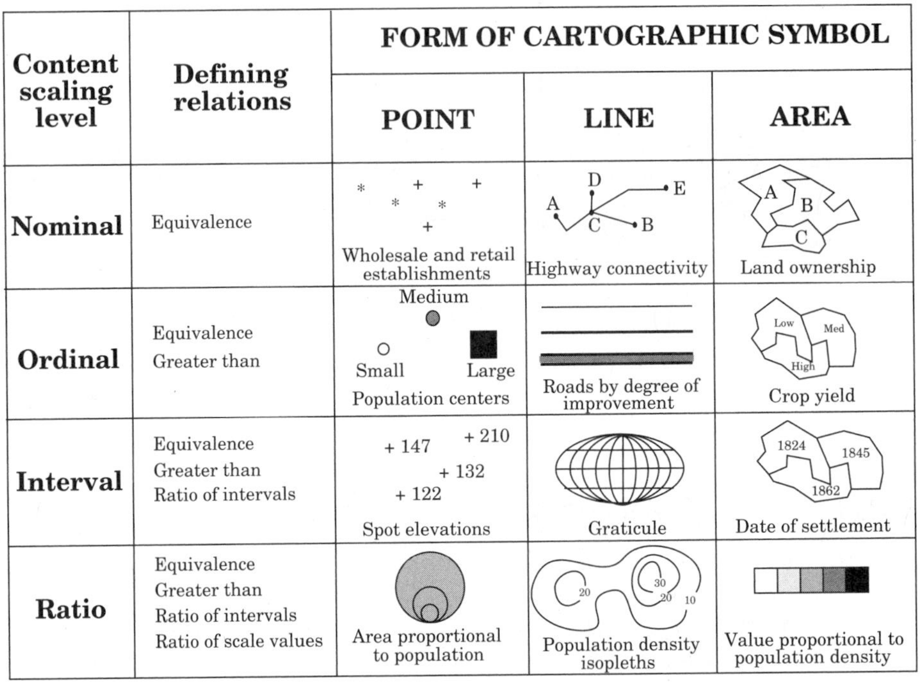

Content scaling level	Defining relations	FORM OF CARTOGRAPHIC SYMBOL		
		POINT	**LINE**	**AREA**
Nominal	Equivalence	Wholesale and retail establishments	Highway connectivity	Land ownership
Ordinal	Equivalence Greater than	Medium Small Large Population centers	Roads by degree of improvement	Crop yield
Interval	Equivalence Greater than Ratio of intervals	+147 +210 +132 +122 Spot elevations	Graticule	Date of settlement
Ratio	Equivalence Greater than Ratio of intervals Ratio of scale values	Area proportional to population	Population density isopleths	Value proportional to population density

Figure 10.1 Classification by scaling and dimension. (After Robinson and Sale, *Elements of Cartography, 3d ed.* ,© 1969, by John Wiley & Sons, Inc. Used with permission.)

conceptual viewpoint on analytical cartography (Tobler, 1979). Tobler's idea of transformations originated with a single set, the transformations of map space caused by map projections, that is, point-to-point transformations. This was expanded to cover other dimensions, and transformations between dimensions. To these can be added transformations of scale and the mapping transformation that actually produces the map, known as the symbolization transformation. In this and the next chapter, these four types of transformations will be examined in detail and used to present a transformational view as a unifying theme in analytical cartography.The idea of transformations in cartography has roots in the work of the cartographer Arthur Robinson, who devised a way of classifying the types of cartographic symbolization and map types in common use (Robinson and Sale, 1969). Robinson placed methods into the framework of a set of states classified by level of measurement and dimension (Figure 10.1). Robinson proceeded to place most of the methods used to make maps into one or other of the categories. David Unwin expanded upon Robinson's classification to add the distinction between map types and data types. Data properties define data types, while the type of symbolization defines map types. This implies that cartography involves a data type-to- map type *symbolization* transformation. Analytical cartography is the scientific treatment of the properties of the data to map type transformation.

To start with Robinson's classification, the primary concern was with the symbolization transformation. The idea was to break down the ways in which symbolization takes place by two factors: the *type of data* and the *level of attribute measurement.* In this context, the symbolization method means those actual symbols used on a given map; blue tints, brown lines, dot patterns, and so forth. The map type is the cartographic technique used to produce the map. As an example, a map with areas shaded red or blue depicting the winning political party within counties has a data type that is area/nominal and a map type that is colored areal. Using red and blue and shading the areas within the boundaries of the counties are the results of the choice of symbolization technique. Under the Robinson/Unwin system, the *states* in the mapping process could be summarized as area/nominal data to colored area map to, say, color ink-jet print from a particular choropleth mapping package. The sequence is *data type* to *map type* to *symbolization* to *map.*

The framework for cartographic transformations proposed here is that the mapping process is the sum of a series of state transformations and their interactions. At any stage in the process, the cartographic information with which we are working has a data structure, and the transformations operate upon this structure. By using Robinson's types of symbolization and Unwin's data and map types, we can lay out a finite set of transformations, each of which can then be examined in more detail. Before we go into the *how* however, we should be fully aware *why* we are performing these important transformations.

10.2 REASONS FOR TRANSFORMING CARTOGRAPHIC DATA

Three of the most important reasons for transforming cartographic data are generalization, converting the geometry of the map base, and changing the data structure. The first reason, *cartographic generalization,* underlies much of analytical cartography. A good subtitle for the discipline of cartography would be a "discipline in reduction" because we are always changing map scales from larger to smaller as we move from data to map and attempting to understand what happens in the process. We may wish to have maps, which have been collected in different ways, transformed to a common basis. It may be a common statistical basis, where we have directly measurable quantities of the same values: where the same statistic, for example, needs to be compared across different countries. Or we may want maps put onto the same geographic reference base. For example, we may want maps with different map projections transformed to one common coordinate system.

Another reason for transformations is the need for converting the geometry of the map base. Cartographic analysis, modeling, different map types, and symbolization methods all favor a particular cartographic data type. If we want to do automated contouring, we somehow have to convert our data into integers in a grid data structure, because this is how about half of the automated contouring packages work. The problems of transforming cartographic data are usually as important as the problems associated with making the final map.

Another major reason for cartographic transformations is to *change data structure.* The different phases that we go through when we handle cartographic data, such as data input or output and data manipulation of various different data types, all have their best structures. The best way to digitize a polygon map for choropleth mapping, unless you want to duplicate information, and assuming that one wants the capability to do some editing of the data after initial entry, is to enter the data as connected line segments forming chains. Then we deal with the problem of transforming them into the format that we really want, which in this case is the polygon list, so that we can shade the areas as required by our symbolization method. Transformation between data structures is worthwhile because the real difficulty is digitizing, and we want to make this stage as easy as possible.

Data structure transformations, however, are for the purpose of suiting the cartographic data for different types of analyses or symbolization and are not an end in themselves. Many other processes involve exactly the same issue; for example, certain analytical operations are far better performed on a grid than on a vector data structure, and vice versa. We almost always find ourselves in need of transforming between data structures, so that the data structure transformation is usually the first, if not the most important transformation, we may apply during the mapping process.

10.3 TYPES OF DATA

The basis of Robinson's classification of symbolization and map types was the division by data dimensions: the familiar division of cartographic objects into the types—*point, line,* and *area*—to which can be added *volume.* Point data can be described simply by a location with an attribute. Line data are usually thought of as being strings of connected straight-line segments, but can also be represented by a mathematical function such as a polynomial or a spline. Area data are regions or polygons, predefined geographic areas with a boundary, a given topology, and an attribute, such as the state of New York. Volumes can represent either continuous statistical distributions, such as measurements of air pollution or terrain elevations.

The second division of cartographic objects under Robinson's classification is by the level of measurement of the data. *Level of measurement* means the classification of the attribute data by its complexity. This division was defined by Stevens (1946). The normal categories we use are *nominal, ordinal, interval,* and *ratio.*

Nominal simply means assigning a name or label. A nominal cartographic feature may be a place name associated with a point object, such as the label **New York** assigned to the location 40 degrees 40 minutes north latitude, 73 degrees 58 minutes west longitude on a small-scale map. Such objects may have rules or methods associated with them, especially in relation to how they are cartographically symbolized and what map types can be used with them.

Ordinal data has a sequence or ranking associated with it, so relations like "larger," "smaller," or "greater than" can be used.

Interval data has a measured numerical value as an associated attribute, such as a measured field elevation. The scale, however, is not absolute but pegged to an arbitrary unit or false zero. Thus a geological map may classify rock types by age before present (BP), representing an age that cannot be precisely determined, yet it may give a relative history to the rock sequences that is less arbitrary than a set of younger than/older than references.

A *ratio* value is a measured value on a scale with a meaningful zero on which absolute mathematical operations can be performed. The Kelvin temperature scale, for example, has a zero determined by the lowest possible temperature rather than the temperature at which ice melts. Examples of ratio values are population density, change rates, and percentages. Transformations between levels of measurement are not necessarily map data transformations, but they do affect the types of maps we can use to portray the transformed data.

A good example, within a normal cartographic representational procedure, is in the making of a choropleth map. We may collect data on population (ratio) and the areas of states (ratio). From these two ratio values we compute a third, in this case population density in persons per square kilometer. This is a simple numerical, mathematical transformation, and it is invertible on the map because we will be showing the outline of the states, and therefore allowing the computation of their areas (if we show a scale). Next we represent our ratio value using an area technique known as choropleth mapping, which first reduces the ratio data to ordinal by classification and then assigns ordinal shade categories and patterns to the regions on the map. The entire set of level-of-measurement transformations is depicted in Figure 10.2. Normally, the symbolization transformation is not an invertible transformation, and frequently it is far from simply performed, because there are many possible ways to subdivide the data into categories.

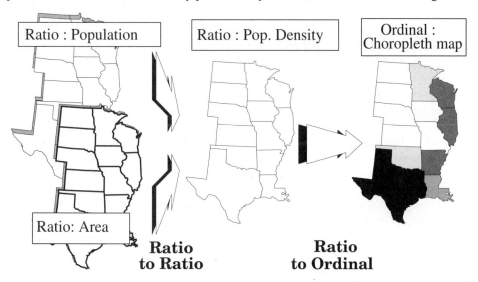

Figure 10.2 Level of measurement transformations for choropleth mapping.

Jenks and Caspall (1971) provided a means of measuring the error involved in this last data transformation, allowing an objective evaluation of the various classification methods. In fact, Jenks and Caspall presented four measures of the accuracy of this transformation. One of these was a composite of the remaining three, which were all based on comparison with an idealized model of a choropleth map with enough classes to have one class per data value. The error was computed using the concept of a three-dimensional solid raised to the height of the choroplethic data value and the standard deviation.

With only one choropleth value (the worst case), the entire map would be represented by the data mean. The overview error would then be the sum of the absolute values of the data values minus the mean, in each case, times the areas represented on the map. Similarly, the tabular error is simply the height differences, without the transformation to volumes. The two versions of this measure are the absolute and relative values of this number. The error indices provide an objective function to maximize, and Jenks and Caspall presented a worked example of a classification optimization for the choropleth data they used. An optimal set of choropleth mapping transformations would minimize this error.

For a choropleth map we can transform the data into a map and simultaneously make definitive statements about the error or information loss in the transformation, which defines the invertibility of the transformation. These types of transformations, their characteristics, how they can be performed, their invertibility, and the error both spatial and aspatial that they embody form the bulk of the subject area for analytical cartography. So familiar are these transformations that we often perform them without thinking, or even perform then mentally while looking at a table of numbers or a map.

10.4 TYPES OF MAPS

David Unwin presented a division of the Robinson classification into data types and map types. Where Robinson was interested in cartographic symbolization, Unwin was interested in the spatial analytic aspects of the state descriptions. Unwin presented two three-by-four tables, shown as Figure 10.3. The classification was the same as Robinson's, with the exception that volumes (surfaces) were added as a data type. The division into map type and data type was based on the distinctions between cartographic entities and their symbolization methods.

Under Unwin's schema, a nominal linear data type was a road, and a nominal linear map type was a network map. The road could be described as the cartographic entity. The network map, however, is both a type of *symbolization* and a type of *cartographic object*. In the following discussion, we will distinguish between the transformation of a cartographic entity to a cartographic object via geocoding and the adoption of a cartographic data structure, and the eventual symbolization transformation of the object required to make a map. The emphasis is not, therefore, on map types or data types, but on the transformations between the states into which cartographic entities, objects, and symbols fall.

The domain of interest of the analytical cartographer is the full set of state transformations possible, yet it is also obvious that there will be an optimal set or pathway of transformations to make a particular map. The symbolization transformation, in which the map is realized and finds a physical description and a tangible reality, forms the topic of computer cartography, being by definition technology dependent. Under the transformational approach, four major cartographic transformations seem to shape the way in which mapping takes place. These are the following: the geocoding transformation between cartographic entities and cartographic objects, involving transitions between levels of measurement, changes in dimension, and changes in data structure; changes in map scale; changes in the locational attributes of the data, that is, transformations of the map base itself; and the transformation resulting in symbolization and a real map.

DATA TYPES

	Point	Line	Area	Volume
Nominal	City	Road	Name of unit	Precipitation or soil type
Ordinal	Large city	Major road	Rich county	Heavy precipitation Good soil
Interval Ratio	Total population	Traffic flow	Per capita income	Precip. in mm Cation exchange

MAP TYPES

	Point	Line	Area	Volume
Nominal	Dot map	Network map	Colored area map	Freely colored map
Ordinal	Symbol map	Ordered network map	Ordered colored map	Ordered chromatic map
Interval Ratio	Graduated symbol map	Flow map	Choropleth map	Contour map

Figure 10.3 Map data and map types. (After Unwin, 1981.)

Finally, we should note that transformations of cartographic objects can yield objects with fewer dimensions or even no dimension at all (scalars or just simply, measurements). The intersection of two lines produces a point, and the measurement of area of a polygon produces a scalar. Cartographic measurement, part of cartometry, is itself simply a transformation to a lesser dimension object or a simpler geometry.

10.5 TRANSFORMATIONS OF THE MAP SCALE

To cartographers, the range of interest is the map scales between about 1:1,000 up to about 1:400,000,000. Remember that *large scale* means a representative fraction with a smaller number as the denominator and covers a small area in a great deal of detail. A 1:1,000 map is approaching a *plan* and is of as much interest to the surveyor or the engineer as to the cartographer. Similarly, a *small-scale* map has a large number in the denominator of the representative fraction and shows a large area in less detail, perhaps a whole state, country, or even a continent.

To provide some examples of these scales, at the large-scale end 1 meter on the ground is 1 millimeter on the map; at the smal- scale end, the entire earth map would fit onto the surface of a gum ball. We have different names for the disciplines concerned with scales beyond these ranges. At extremely small scales, the concerned disciplines are astronomy and planetary physics; at extremely large scales, the fields include planning, surveying, and engineering. The scale of one-to-one is called *reality*, and scales beyond this are the enlargement disciplines of chemistry, microbiology, and particle physics. We use different instruments at different scales, and the scales affect the collection of data with those instruments. We cannot, for example, measure interplanetary distances with a ruler; neither can we observe distant galaxies with an electron scanning microscope.

Cartographically, transformations between scales are almost always from larger to smaller scale and proceed by a process of *generalization,* involving the selection of features to survive the scale change, the deletion of detail, the smoothing or rounding of the data, and sometimes the change in symbolization necessary to deal with the new scale, which can even involve moving objects to facilitate interpretation. Some automated methods for scale modification are discussed in Leberl et al. (1986). Figure 10.4 summarizes the types of scale change discussed by Weibel (1987) in the specific context of digital elevation models but applicable to most types of cartographic data. Weibel noted that cartographers perform four basic operations while generalizing cartographic data. These are elimination (selection), simplification, combination, and occasionally displacement.

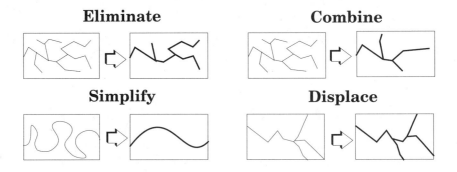

Figure 10.4 Scale change transformations

Elimination consists of dropping cartographic features as we move from large- to small-scale maps. *Simplification* involves classical generalization, such as placing a dot for a city on a small-scale map or showing a river as a smoother and smoother line at smaller and smaller scales. *Combination* involves joining features together at smaller scales, such as joining a river with its tributaries or combining island groups into a smaller number of "representative islands." *Displacement* is occasionally necessary to reveal structure or to allow space for labels. This is often the case when many features are clustered together in a single region on the map.

Our scientific ideal is always to collect data at scales that are larger than we need for a particular map so that we can eliminate, simplify, and generalize using known rules, methods, and criteria. Much of manual cartography, and a great deal of computer cartography, has to do with sets of rules for determining the generalization transformation. Rules and algorithms for scale-related generalization have been researched in considerable detail in recent years (Buttenfield and McMaster, 1991; McMaster and Shea, 1992).

Most algorithms perform the simplification stage, that is, they reduce the number of points in a digital cartographic line. Algorithms include retaining only every nth. point, selecting points based on the line's interior angle, selecting points outside a buffer of given width following the line (epsilon filtering), and selecting points recursively based on the maximum orthogonal distance from the line above a threshold.

Different map lines have distinctly different properties, because they reflect different processes on the ground. Map administrative lines are straight or follow meridians and parallels. Contours are smooth but have sharp breaks at streams, and coastlines sometimes show scale-independence or fractal characteristics. Probably the best defined are the rules for generalizing a coastline. We generalize through scales, gently smoothing the coastlines, but we also make jumps at certain scales. When an island becomes a dot at a particular scale, we may choose to eliminate it. Similarly, whole inlets or river estuaries may disappear on small-scale maps. Figure 10.5 shows this phenomenon. The lake in question actually appears to change shape as it moves though the scales at which it is mapped.

Only recently have cartographers thought about the inverse transformation between scales. This is probably because we used to call this process "cartographic license" or sometimes "witchcraft." A better name would be *enhancement,* and it is important to note that not all enhancement is witchcraft, for we may have good information at some places on the map about patterns at larger scales.

Extending the characteristics of this variation uniformly over the map artificially may be termed *emphatic enhancement.* On the other hand, using a model to generate variation, such as fractal enhancement of coastlines, is *synthetic enhancement.* Dutton (1981) (Figure 10.6) was a pioneer in this work and his ideas have since been extended to other types of data (Clarke et al., 1993).

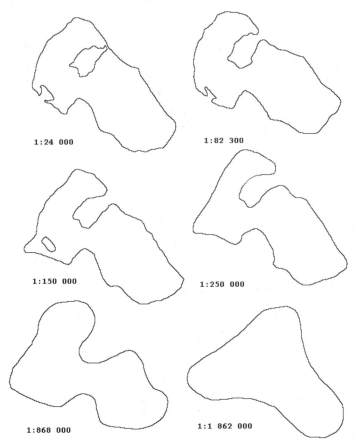

1:24 000

1:82 300

1:150 000

1:250 000

1:868 000

1:1 862 000

Figure 10.5 Copake Lake (in New York State) at different map scales.

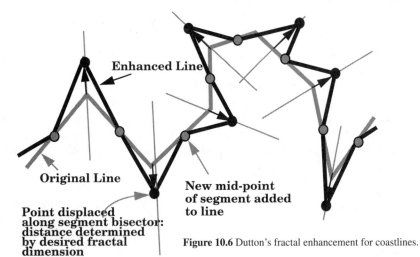

Enhanced Line

Original Line

Point displaced
along segment bisector:
distance determined
by desired fractal
dimension

New mid-point
of segment added
to line

Figure 10.6 Dutton's fractal enhancement for coastlines.

10.6 TRANSFORMATIONS AND ALGORITHMS

The most elegant statement of the characteristics of a transformation is in a mathematical equation. Invertibility of equations can be proven by algebra or other methods, so that we can rearrange them to find other variables given a different set of starting conditions. When dealing with cartographic transformations, we do indeed have many transformations that are expressible in the language of mathematics. Others, however, are less amenable to such simple description.

In computer science, an *algorithm* is a special method of solving a problem. An algorithm may be stated either as a simple formula or as a set of sequential instructions to be followed to arrive at a solution. Church's theorem implies that if a process can be so stated, then it can be automated, in this case programmed. The set of algorithms for cartographic transformations operates on cartographic data that have some preexisting data structure, and results in a form of the data that is closer to the map required. A good summary, with apologies to Wirth, would be that

```
data structures + transformational algorithms = maps
```

When you dig down far enough within any piece of cartographic software, however complex, cartographic transformational algorithms are the very nuts and bolts with which that software is constructed. Similarly, the cartographic data structures involved are the materials that the transformations hold together to produce maps. An understanding of the data, the problems, the data structures, and the algorithms is really all the analytical cartographer needs, except of course a real desire to make *better* maps. In the following chapters of Part III on analytical cartography, we will go closer into the algorithms that perform cartographic transformations, and show how some simple computer programs can be written to implement the transformational approach taken here.

10.7 REFERENCES

Buttenfield, B. P. and R. B. McMaster, eds. (1991). *Map Generalization: Making Rules for Knowledge Representation.* New York: Wiley.

Clarke, K. C., R. Cippoletti, and G. Olsen (1993). "Empirical Comparison of Two Line Enhancement Methods." *Proceedings, AUTOCARTO 11,* Eleventh International Symposium on Computer-Assisted Cartography, Minneapolis, pp. 72–81.

Dutton, G. H. (1981). "Fractal Enhancement of Cartographic Line Detail." *American Cartographer,* vol. 30, no. 1, pp. 23–40.

Jenks, G. F., and F. C. Caspall (1971). "Error on Choroplethic Maps: Definition, Measurement, Reduction." *Annals of the Association of American Geographers,* vol. 61, no. 2, pp. 217–244.

Leberl, F. W., D. Olson, and W. Lichtner (1986). "ASTRA—A System for Automated Scale Transition." *Photogrammetric Engineering and Remote Sensing,* vol. 52, no. 2, pp. 251–258.

McMaster, R. B., and K. S. Shea (1992). *Generalization in Digital Cartography.* Association of American Geographers Resource Publication Series, Washington, DC.

Robinson, A. H., and R. D. Sale (1969). *Elements of Cartography,* 3d ed. New York: Wiley.

Stevens, S. S. (1946). "On the Theory of Scales of Measurement." *Science,* vol. 103, pp. 677–680.

Tobler, W. R. (1979). "A Transformational View of Cartography." *American Cartographer,* vol. 6, no. 2, pp. 101–106.

Unwin, D. (1981). *Introductory Spatial Analysis.* London: Methuen.

Weibel, R. (1987). "An Adaptive Methodology for Automated Relief Generalization." *Proceedings, AUTOCARTO 8,* Eighth International Symposium on Computer-Assisted Cartography, Baltimore, pp. 42–49.

11

Map Transformations

11.1 TRANSFORMATIONS OF OBJECT DIMENSION

In the Chapter 10, we met the transformational view of cartography. Two types of transformations were considered: first, transformations between the types of data and types of map involved in the mapping process; and second, transformations between map scales. A third type of transformation is the transformation of cartographic objects themselves via their dimensions. Because these are transformations of either the locational geometry of the data or of the structure used to represent a cartographic object spatially, they deal directly with the core of map information and are close to the heart of analytical cartography.

We have already noted the contribution of Tobler's transformational view (Tobler, 1979). Tobler saw cartographic object dimensions as states for cartographic data, and he enlarged the set of state transformations beyond the classical point-based cartographic conversions to include the other dimensions with which we specify locations. He envisaged a three-by-three table—of points, lines, and areas by points, lines, and areas— that contained nine possible groups of transforms, with many specific cases and divisions in each category. With volume added as a cartographic measurement dimension, we have 16 possible sets of transformations, which can be arranged in a four-by-four table. Tobler saw this table as a transformation matrix, a set of possible transformations between states at different times. Within this viewpoint, time 0 and time 1 are states, and the cartographic data go through a transformation between states (Figure 11.1).

Changes within this transformation matrix along the diagonal are transformations either between scales or between data structures. The final transformation as far as map production is concerned is that between the cartographic data and a map, or the symbolization transformation. This chapter will focus upon the dimensional transformation, and finish by discussing the symbolization transformation. Chapter 12 follows up specifically with a more in-depth discussion of scale transformations and data structure conversions.

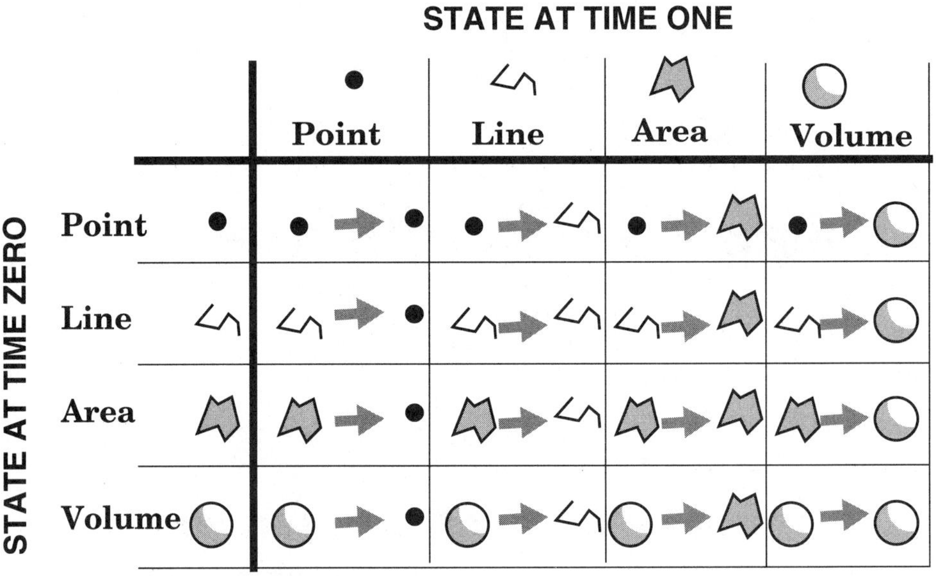

STATE AT TIME ONE

Figure 11.1 The 16 possible single step dimensional transformations.

Using matrix algebra to express transformations, we can express a sequence of transformations as matrix multiplication transformations of a vector of (x, y) coordinates, X:

$$X = \begin{bmatrix} x_1 & y_1 \\ x_2 & y_2 \\ x_n & y_n \end{bmatrix}$$

$$XT = X' \qquad \text{Normal transformation.}$$

$$X'T^{-1} = X + E \qquad \text{Inverse transformation.}$$

$$TT^{-1} = E \qquad TT^{-1} = I \qquad \text{Transformation both with and without error.}$$

Under normal circumstances, the cartographic transformation is imperfectly invertible, due to error in the transformation process. The transformation, times its own inverse, gives not the ideal identity matrix (I), but a vector of errors E. Analytical cartography seeks to identify, model, eliminate, or reduce the matrix E so that its spatial effects are known and measureable or negligible.

What we would like to be able to do as analytical cartographers is to express a cartographic transformation either as an explicit mathematical operation such as the matrix formulations above, or as an algorithm that allows us to project state 0 onto state 1 in a fully described way. The transformation from state 0 to state 1 has an inverse transformation that transforms from state 1 to state 2, equivalent to state 0. If such a transformation exists, the entire transformation belongs to a special subset of transformations known as invertible transformations. Knowledge of invertible transformations allows prediction of error and its detection and elimination, understanding of the underlying geographic phenomena, and spatial modeling of the real processes behind the cartographic data.

In this chapter, we shall deal with transformations at the low-dimensional end of the transformation matrix, especially point-to-point transformations. Most of these are transformations within a single type but between scales, and of attribute data scaling, usually known as map generalization. Some of them have as their input a cartographic line and as their output a measurement based upon properties of the line. Not all are invertible transformations.

Point-to-point transformations, which Tobler called the classical cartographic transformations, are very central to analytical cartography. An understanding of point-to-point transformations is essential to analytical cartography, more essential than many of the other dimensional transformations. This is because most higher-dimensional cartographic objects can be reduced to a set of locational measurements at points.

Map-based transformations are pivotal to understanding the ability of analytical cartography to make systematic inquiries into how data structures representing cartographic objects encode and represent space. Analytical cartography seeks to determine how the geographic properties of space sharing can be used in analysis, modeling, and prediction.

The transformation matrix shown in Figure 11.1 shows only the single-step transformations. It is possible also to have multiple-step transformations, and these are indeed common. As an example, if we measure elevations in the field, it is most convenient to record them at sets of irregularly distributed points. We can describe these points digitally as point locations with the attribute of elevation.

The data value we have sampled, topography, is volumetric, we have sampled only the upper surface with sets of point data observations. For example, we could take a continuous value, distributed over a volume surface, sample the value using a series of points with attribute values associated with them, and then transform the attributes of the points over space using a TIN data structure (Figure 11.2).

Our motive may be, perhaps, that we want to use an automated contouring technique requiring data in the TIN structure. We transform from state 0 to state 1, and then on to state 2. State 1 to state 2 is a point-to-area transformation, where the first set of points is irregularly spaced and the set maps onto a set of triangles forming a triangulated irregular network. We might take the set of triangles and use it to feed a set of lines through the area to make contour lines. So we have input a series of points, transformed them to areas, and then transformed them to a set of lines. We then draw the lines on the map as contours and use them in the map reader transformation to communicate the impression of topography, the upper surface of a volume.

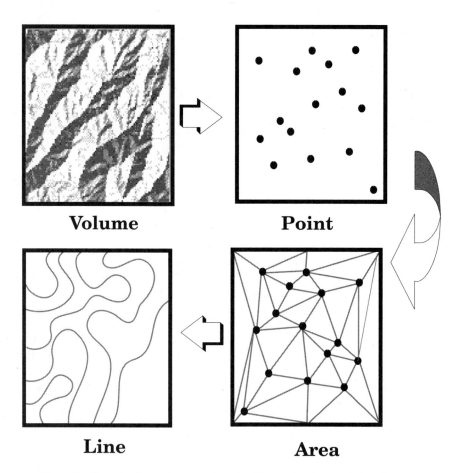

Figure 11.2 Example of a multistep transformation. Volume (terrain) sampled at points, converted to TIN (areas) then contoured (lines).

Using this transformational view of cartography, we have a versatile framework for analytical cartography, and we can place any kind of cartographic data manipulation into this framework. In the remainder of this chapter, we will take one set of transformations and start working through specific examples. These are transformations of the *location measurements* themselves, in other words, coordinate transformations. This type of transformation changes the geographic space of the base map itself. For all maps, but especially small-scale maps, with such transformations we are mainly talking about *map projection* transformations.

11.2 MAP PROJECTION TRANSFORMATIONS

We can have three approaches to transforming the coordinates of small-scale maps. The simplest approach is to ignore the problem of the earth's shape, and this is a frequently employed solution to the problem. If the area of mapping interest is small, such as an area that could be paced or taped out, or surveyed across a construction site, then the simple plane is a reasonable approximation of the geometry of the object to be mapped.

Even at large scales, however, ignoring the projection results in some unexpected, inaccurate spatial statistics, and in poor maps, especially if we take maps that have been prepared on different map projections and then overlay them. Simply mapping 1-degree by 1-degree cells onto a single display square seems to work fine at low and middle latitudes, but as we approach the poles, the north-south distortion becomes sufficient to make computations of areas, directions, and bearings meaningless.

The second approach to coordinate transformations is to use statistical techniques such as rubber sheeting or adjustment. Surveyors have applied adjustments to locational measurements for many years, distributing the random error of measurement proportionately around points and along transects. In remote sensing, images are transformed into map space using the same principle, that of a least squares fit of the image to the map geometry. Under ideal circumstances we would know every geometric property of the map and understand the relationship of these geometric properties to latitude and longitude. This would allow us to know all the equations to perform the space transformation. This is rarely the case.

Rubber sheeting is a statistical and empirical approach. It works, it is frequently used, and it is acceptable cartographically as long as we realize that we have incorporated systematic error into our final map and that we have lost some of the cartographic fidelity. Of course, because this method is empirical the transformation is not directly invertible, although it often comes close to it if we retain the control point locations.

The third approach is to compute the characteristics of the transformation itself geometrically. There are two ways of doing this; we can be precise, or we can be approximate. Which we choose depends largely upon what we want to do with the data. For example, we can choose any one of three geometric forms to model the earth's surface: a sphere, an oblate ellipsoid, or a geoid.

The sphere is most familiar to us as the common globe and is perhaps the easiest to understand. Numerous estimates of the radius of the sphere have been made, and are used in mapping specific areas. The oblate ellipsoid is a more precise model, allowing higher levels of accuracy. An oblate ellipsoid is the volume traced out by an ellipse rotated about its minor axis. Because the earth is about 42 kilometers fatter across the equator than it is pole to pole, the difference from a sphere is small.

A common estimate of the degree of flattening is 1:294.98, an estimate established in 1866 by Alexander Ross Clarke and still normally used for the mapping of the United States, although it will be phased out as the 1983 North American Datum (NAD83) becomes the standard. Because it was established that different figures could be computed for different parts of the globe, a "best-fit" ellipsoid was chosen and is given as the 1980 Geodetic Reference System value of 1:298.257, the basis of the NAD83 (Snyder, 1983).

The ellipsoid is important, because when maps are referenced to it directly they are more accurate and capable of higher precision. The third earth model is even more precise. In this model, we fit an empirical surface to the earth's gravity field, which follows mountain ranges and even dips below and above sea level. How far this deviates from the ellipsoid, or usually the local best-fit ellipsoid, depends on where we are, giving us a lumpy ellipsoid-like figure known as a geoid. How precise we need to get depends on our function. A map of U.S. population clearly does not need to be precise beyond the best-fit sphere, while the guidance system of a spacecraft needs much more accurate and precise calculations.

Cartography traditionally has worked with the sphere because the manual construction of map projections is tedious. The sphere is simplistic, but it makes the calculations and the construction of projections using drafting instruments much easier. Map projections usually can be described as projections from a point source on a sphere to points on a flat plane. Map projections involve two different types of mappings, both of which benefit from use of the computer. Figure 11.3 shows a representation of a sphere and the various types of possible mappings of points on the sphere onto points on a map.

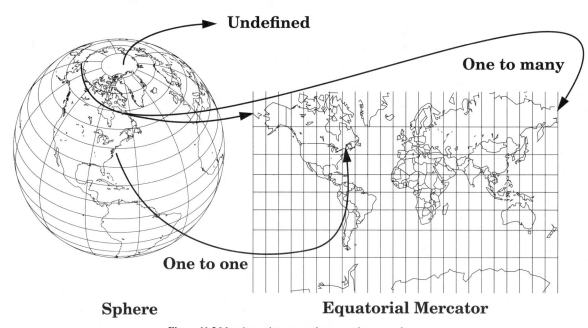

Figure 11.3 Mapping points on a sphere to points on a plane.

Each point on the earth's surface maps onto a single point on the map. Geometry actually uses the term *mapping* to describe this process, borrowed from cartography's ancient Greek roots. This is usually called a *one-to-one mapping,* and it is preferred because it is well behaved and simply invertible. Imagine an old favorite, Gerhardus Mercator's projection of the earth, one we are all familiar with from the wall of our elementary school classroom.

Think about the usual dividing line on the equatorial Mercator projection, 180 degrees east (or west). This single line does not map onto a line; it maps onto two lines and becomes undefined as we approach the poles. This is an example of a *one-to-many* mapping, and an *undefined* mapping, and these are the kinds of mapping transformations that are extremely difficult to invert.

Even when the projection is one-to-one, the cartographer is faced with the dilemma of portraying a spherical or ellipsoidal earth on a flat surface. In general, projections are either *conformal,* in which they preserve local shape and direction at locations on the map, or *equivalent,* in which they preserve area. To achieve one or the other, projections are sometimes cut or interrupted, usually along parallels and meridians. Often a projection is neither conformal nor equivalent, but a compromise between these properties. The projection should therefore be taken into account carefully when one is using cartographic algorithms such as area computation or angle calculation on the geographic coordinates. Different geometries apply in the spherical and planar cases. For example, most people are familiar with the geometry of plane triangles (Figure 11.4). We know things like the law of sines. If a triangle has sides A, B, and C, with opposite angles a, b, and c, then we know that

$$\frac{\sin A}{a} = \frac{\sin B}{b} = \frac{\sin C}{c}$$

We often use this and other formulas to solve general-purpose triangles. It is a little different on a sphere where we use the spherical triangle. For example, if you are flying an airplane from New York to Chicago, you may want to minimize the amount of fuel you use by flying along the shortest path. To fly along the shortest, or *great circle,* route, you would have to fly along the shortest distance on the surface of a sphere between two points. This is a different geometry from the regular triangular case. We have a spherical

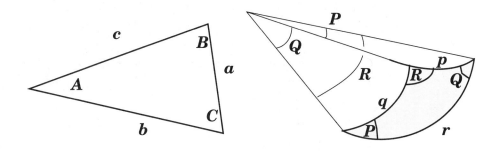

Figure 11.4 Planar and spherical triangles.

rule of sines, because the lengths of our sides are angles, too. If lengths are angles, they

$$\frac{\sin p}{\sin P} = \frac{\sin q}{\sin Q} = \frac{\sin r}{\sin R}$$

are usually expressed in radians rather than in degrees, because radians are used by the computational functions involved. We usually express latitudes and longitudes in degrees minutes and seconds, but in most spherical geometry we deal with radians, and most of the angles are angles at the center of the earth. In particular, computer programs and calculators often require angles in radians. Under the Spatial Data Transfer Standard, the required use is the decimal degrees format. Among existing software, the degrees, minutes, seconds format, that is, DD.MMSS, is frequently encountered.

To convert angles in degrees to radians, use the following formula:

$$\text{radians} = \frac{\pi}{180} \times \left(\text{degrees} + \frac{\text{minutes}}{60} + \frac{\text{seconds}}{60 \times 60} \right)$$

Geodesy works with several alternative means for representing earth coordinates, including three-dimensional (x, y, z), polar coordinates, three angle coseries, and the familiar latitude and longitude.

Let us now set up a spherical angular referencing system based on latitude and longitude. You will notice that this involves going through a transformation so that you can get a range of values. Figure 11.5 shows a longitude that is an x value going from negative 180 at the international date line (or more strictly, the 180th meridian) to positive 180 at the international date line. Notice that at the dividing line we have a one-to-many mapping. We have chosen a central meridian, which in this case happens to be the origin of

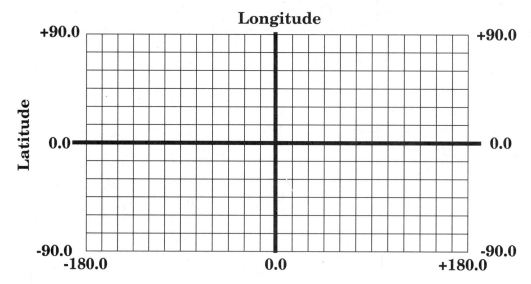

Figure 11.5 Generalized world coordinate system.

zero at the prime meridian. Going up and down, we choose the equator as zero; we make the North Pole 90 and the South Pole -90, that is, 90 degrees north and 90 degrees south. Notice that here is another one-to-many mapping; that is, that the coordinate system as shown is inherently a cylindrical projection, or more precisely, an equirectangular projection. On this map we have one-to-many mappings around the edge, and one-to-ones within the map itself, where fortunately we can handle most of the transformations without too much trouble.

Any map projection transformation starts with latitude and longitude pairs and yields Euclidean coordinates (x, y). With the correct scale factor, the x and y values are measurements in map millimeters which we can use in a graphics program when we plot a map. λ and ϕ, longitude and latitude, respectively, are given in radians.

Map projection transformations deserve a closer look here before we move on to planar-only transformations. The Mercator projection, which strictly we should call the equatorial Mercator projection, is a cylindrical projection. It is based on a projection of the Earth's surface onto a cylinder wrapped around the earth so that it just touches at the equator and has a central axis which passes through both poles. The formulas for Mercator's projection are

$$x' = Rs\,(\lambda - \lambda_0)$$

$$y' = Rs\log\left(\tan\left[\frac{\pi}{4} + \frac{\phi}{2}\right]\right)$$

where R is the radius of the sphere on the map, s is the scale factor (representative fraction of the map), and λ_0 is the longitude in radians of the central meridian. Because the radius of the earth at the equator (under the Geodetic Reference System of 1980) is 6,378,137 meters, scale factors have to reduce that to some reasonable value so that the projected map will fit onto a computer screen or a sheet of plotter paper. At 1:42,000,000 the earth can be represented by a globe about as big as a basketball. With s set to 0.00000001, the map will have an equatorial diameter of about 0.4 meter.

In most Mercator projections the origin is the prime meridian (longitude zero) so most Mercator projections appear with Europe in the middle, or at least with Greenwich in the middle, as this is the origin of the coordinate system. There is nothing that says we have to focus the projection there, but it is rather convenient, because the international date line becomes the space and time dividing line for the map. Also, the one-to-many mapping at the date line is least inconvenient; it passes mostly through water. Strictly speaking, the 180th meridian is not the international date line, because the line is moved to avoid land. We can plot world maps on the equatorial Mercator projection using the transformation equation. The spacings on the Mercator projection latitude graticule get closer and closer as we approach the equator, the effect of taking the logarithm.

Map projection is a part of analytical cartography where the use of computers is quite advantageous because the repetitive computations are very tedious. Most people can work through the formulas here, but it is indeed humbling to realize that Gerhardus Mercator presented the projection as a graphic construction in 1569 and that the formulas were published in 1599 by Edward Wright. This was a significant moment in the history of cartography, showing the two approaches of a technology-dependent graphic (using Mercator's pen, paper, compasses, and dividers) and Wright's analytical cartographic approach (using mathematics). One wonders whether the mathematical approach presupposes and requires the visualization inherent in the graphic.

Function 11.1 is a C language function designed to perform the equatorial Mercator transformation for all the points in a string stored in the structure STRING, introduced in Chapter 7. Note that when the longitude of the origin is not zero, the programmer must rotate the data between −180 and +180.

The same transformation can also be expressed using matrix notation. If we start with longitude and latitude and apply a transformation to them, we get one from the other:

$$\begin{bmatrix} x \\ y \end{bmatrix} = T \begin{bmatrix} \lambda \\ \phi \end{bmatrix}$$

It also works the other way around:

$$\begin{bmatrix} \lambda \\ \phi \end{bmatrix} = T^{-1} \begin{bmatrix} x \\ y \end{bmatrix}$$

If the purpose is to get longitude and latitude, we can start with their cartesian coordinates taken by measurement from a map on some map projection and we can apply the inverse of this transformation to yield longitude and latitude.

Computer cartographic software to perform the transformations between map projections is increasingly part of the more sophisticated computer mapping packages. Especially important is the ability to transform between coordinate systems based on specific map projections, such as the Universal Transverse Mercator and the state plane systems. An excellent reference on map projections is Snyder's *Map Projections—A Working Manual (*Snyder, 1987). This work includes references to computer programs to perform the map projections listed and discussed. The reader should also check Snyder and Voxland's (1989) *Album of Map Projections,* which shows plots of each of the map projections along with their formulas.

The MicroCAM software package is an excellent tool for investigating the characteristics of map projections. MicroCAM computes and plots maps on IBM-PC compatible microcomputers for any of some 27 map projections, including graticules, coastlines, rivers, political boundaries, and United States state and county outlines (Anderson, 1994). The package, written by Scott Loomer, is available from the Microcomputer Specialty Group of the Association of American Geographers at a very low cost.

Also of interest is the NOAA *General Cartographic Transformations Package,* a set of FORTRAN programs and subroutines, along with data files containing all the necessary constants for the various standards and datums as well as constants for the map projections. Twenty map projections, with forward and inverse calculations, are included in the public domain package (NOAA, 1988).

Function 11.1

```
/* Function to transform a string to Mercator Projection */
string #include <math.h>
#define PI 3.141592654
#define R 6378137
#define S 0.00000005
#define LAMBDA_SUB_ZERO 0
STRING mercator(line, max_latitude)
      STRING line;
      int max_latitude;
{

      STRING project;
      double two_pi_over_360, pi_over_four, r_times_s;
      int i, k;
      two_pi_over_360 = 2.0 * PI / 360.0;
      pi_over_four = (double) PI / 4.0;
      r_times_s = (double) R *S; k = 0;
      for (i = 0; i < line.number_of_points; i++) {
      /* Convert to radians */
      line.point[i].x = two_pi_over_360 * line.point[i].x;
      line.point[i].y = two_pi_over_360 * line.point[i].y;
      /* test to exclude points outside range of latitude */
            if((line.point[i].y <= max_latitude) &&
            (line.point[i].y >= -1.0 * max_latitude)) {
                  project.point[k].x = line.point[i].x;
                  project.point[k].y = line.point[i].y;
                  k++; }
      }
      project.number_of_points = k;
      for (i = 0; i < project.number_of_points; i++) {
      project.point[i].x = r_times_s *
                  (project.point[i].x - LAMBDA_SUB_ZERO);
      project.point[i].y = r_times_s * log(tan(pi_over_four
            + (project.point[i].y / 2.0)));
      }
      return (project);
}
```

11.3 PLANAR MAP TRANSFORMATIONS

In the remainder of this chapter, we will consider some simple geometric transformations, still using map coordinates. Even though some of these transformations are simple, they are nevertheless powerful. Generally, we will be dealing here with transformations between point cartographic objects and objects with the same or higher dimensions. In most cases, we will discuss a problem, note the mathematical expression required for a solution, and then provide a computer program to yield the answer. Two types of results are derived: (1) transformations that produce a graphic object, and (2) transformations that produce a scalar measure. The latter can be thought of as a space-collapsing transformation, or a geographic measurement.

11.3.1 Transformations Based on Points

Distance between Two Points

Starting with something point-based and simple, most elementary is computation of the distance between two points. For example, consider two locations in space; call the first (x_1, y_1) and the second (x_2, y_2). Given their locational attributes, otherwise known as their coordinates in space, and assuming planar and not spherical geometry (although spherical distance is no less difficult to compute), what is the distance between them? This is not a puzzling problem, having been figured out over 2,400 years ago by Pythagoras. The distance between those two points is equal to the square root of the sum of the squared lengths of the sides of the triangle made between these two points and the orthogonal axes. In other words, the distance is the square root of the difference between the two x values squared, plus the difference between the two y values squared. Pythagoras actually would have said that the areas of squares formed on the sides on the triangle made by the two points and a right angle with the axes are such that the largest triangle has an area equal to that of the sum of the smaller two.

To compute the distance between two points, which is simply a derived measure:

$$d_{2\text{to}1} = \sqrt{(x_2 - x_1)^2 + (y_2 - y_1)^2}$$

Analytically, the cartographer may use the computed distance to achieve an objective, for example to minimize distance (Function 11.2).

Function 11.2

```
/* Function distance: distance between two points */
#include <math.h>
float distance(one, two)
POINT one, two;
{
        return ((float) pow((two.x - one.x) * (two.x - one.x)
        + (two.y - one.y) * (two.y - one.y), 0.5));
}
```

We may have a whole series of points, and we want to find which two points are closest together. We have to calculate this value for all pairs of points and then sort the values to find the lowest, which will give us the nearest neighbors. We have already formulated an algorithm for distance computation. Distance minimization is a potential application of this algorithm. We can think of a matrix of points by points, with matrix values as the distances between the points, as being a useful transformation of the map space. In this case, the transformation is invertible, because the surveying practice of trilateration allows us to recompute relative locations from distances.

Length of a Line

Length is very similar to distance. A long line in computer cartography usually consists of a string of joined line segments, each consisting in turn of pairs of points with a connecting straight line. The length of each of these segments can be computed using the algorithm above. We simply have to add the lengths of all the segments to compute the length:

$$\text{length} = \sum_{i=1}^{npts} \sqrt{(x_i - x_{i-1})^2 + (y_i - y_{i-1})^2}$$

Function 11.3 uses a line stored in our predefined STRING structure (Chapter 7). Notice that if the line has a number of points in the structure of type STRING given by the value line.number_of_points points, we have line.number_of_points - 1 segments.

Function 11.3

```
/* Function line_length: Return length of a line */
#include "cart_obj.h"
float line_length (line)
STRING line;
{
float distance(), sigma = 0.0;
int i;
for (i = 1; i < line.number_of_points;i++)
     sigma +=distance(&line.point[i],&line.point[i-1]);
return (sigma);
}
```

Centroids

One transformation of point data that yields a point rather than a scalar measurement is the computation of the centroid. Let us say that every city in the United States has a location, (x, y), and also a population. For cartographic reasons, we may wish to associate the population number with the point, to draw a circle in proportional circle mapping, for example. A fairly simple concept is the center of gravity of the population. This single point is representative of a whole scatter of other points, so this is a points-to-point transformation. The Census Bureau usually publishes maps showing how this centroid has moved over time in the United States. How do we calculate this average or descriptive location? Locations in geographic space, perhaps the United States, have eastings (x), northings (y), and values, perhaps populations P. Each instance of the data will be called the nth, and there are a total of npts of them. Assuming planar geometry, we can calculate the average x, $\bar{x}$, the average y, $\bar{y}$, weighted by their populations.

$$\bar{x} = \frac{\sum\limits_{i=1}^{npts} P_i x_i}{\sum\limits_{i=1}^{npts} P_i} \qquad \bar{y} = \frac{\sum\limits_{i=1}^{npts} P_i y_i}{\sum\limits_{i=1}^{npts} P_i}$$

For example, if a city has a population of 12,000, we treat this as 12,000 cities at the same place with a population of only one person.

Standard Distance

Above we computed the mean x and y as an aggregate measure of location, so we can similarly compute the standard deviations in x and y as a measure of dispersal. The standard deviations in x and y can be used to compute the standard distance, a measure of the degree of scattering of point distributions around a focal point, the mean center, in both x and y as s_x and s_y.

$$s_x = \sqrt{\frac{\sum\limits_{i=1}^{npts} (x_i - \bar{x})^2}{npts}} \qquad s_y = \sqrt{\frac{\sum\limits_{i=1}^{npts} (y - \bar{y})^2}{npts}}$$

The root of the sums of these values squared gives a value called the standard distance s:

$$s = \sqrt{(s_x^2 + s_y^2)}$$

This value is a measure of dispersal for point distributions. Now we have a simple measure of aggregate location and dispersal for point distributions. These values can be used analytically, for example, to examine the assumptions surrounding interpolation to a grid.

Nearest-Neighbor Statistic

Another descriptive statistic of point distributions is the nearest-neighbor statistic (*NNS*). For npts points (x, y) scattered over an area A,

$$NNS = 2 \times \frac{\sum_{i=1}^{npts} d_i}{npts \times \sqrt{\dfrac{A}{npts}}}$$

where d is the distance from each point to its closest neighbor. In this case, the point distribution has an empirically derived scalar statistic that measures the resemblance of the point distribution to a clustered, a random, a grid, or a hexagonal point distribution, although the value is critically sensitive to the area used (Figure 11.6).

A C language function to compute this statistic, therefore, must also choose how to compute the area that encloses the points. This can be the area of an enclosing polygon, such as a county or field boundary, or a bounding rectangle can be used. In Function 11.4 code function, the area is passed from the calling program

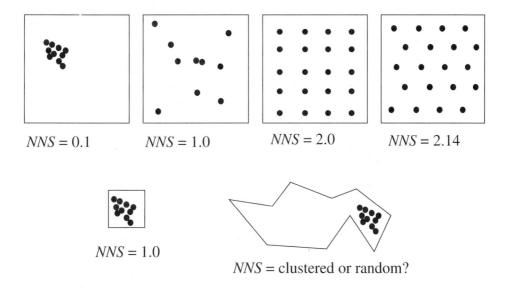

| $NNS = 0.1$ | $NNS = 1.0$ | $NNS = 2.0$ | $NNS = 2.14$ |

$NNS = 1.0$

$NNS =$ clustered or random?

Figure 11.6 Nearest neighbor values and area sensitivity.

Function 11.4

```
/* Function nearest: Nearest neighbor statistic
 * assumes data are in structure POINT.
 * Calling function should provide area.
 * kcc 10-88 revised 8-93 */
#include "cart_obj.h"
#include <math.h>
float           nearest(n_point, point, area)
int             n_point;
POINT  point[];
float  area;
{
float distance(),this_distance,shortest_distance,
      sigma = 0.0;
int             i, j;
for (i = 1; i < n_point; i++) {
      shortest_distance = distance(point[0] point[i]);
      for (j = 0; j < n_point; j++) {
            if (i != j) {
            this_distance =
            distance(point[i], point[j]);
                  if (this_distance < shortest_distance)
                     shortest_distance = this_distance;
                  }
            }
            sigma += shortest_distance;
      }
      return ((2.0 * sigma /(n_point * sqrt(area / (double)
            n_point)))));
}
```

11.3.2 Transformations Based on Lines

Intersection Point of Two Lines

Finding the intersection point of two lines is an example of an algorithm that performs a cartographic transformation yielding a point. A critical use of this algorithm in analytical and computer cartography is in clipping and computing the overlap between sets of chain encoded areas or the intersection of line features. Line intersection computations form an essential component of the extra computations necessary to structure data topologically in mapping and GIS systems. In these systems, the significance of the chain as a basic object is foremost.

This algorithm falls into the category of a test, because in computer cartography we frequently have to test lines for intersection. If two lines intersect, where do they intersect? This computation of line intersection points is an essential part of many vector mode systems and lies at the basis of all clipping algorithms. For the two line segments shown in figure 11.7, we wish to compute the intersection point (p, q). We can find the intersection point by solving two simultaneous equations for the line $y = a + bx$, with two known values. If (x_1, y_1) and (x_2, y_2) lie on the same line, then

$$y_1 = a_1 + b_1 x_1$$

$$y_2 = a_1 + b_1 x_2$$

Similarly, if (x_3, y_3) and (x_4, y_4) lie on the same line, then

$$y_3 = a_2 + b_2 x_3$$

$$y_4 = a_2 + b_2 x_4$$

If there exists an intersection point, (p, q) that lies on both lines, then

$$q = a_1 + b_1 p$$

$$q = a_2 + b_2 p$$

By subtracting the former from the latter,

$$q - q = a_1 - a_2 + p(b_1 - b_2)$$

and rearranging, we obtain

$$a_1 - a_2 = p(b_2 - b_1)$$

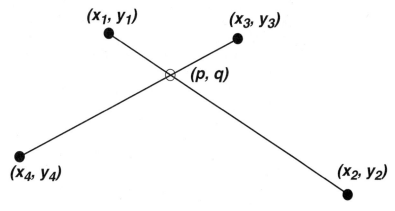

Figure 11.7 Terminology for line intersection.

If an intersection exists, this implies that

$$p = \frac{a_1 - a_2}{b_2 - b_1}$$

and by substitution

$$q = a_1 + b_1 \frac{(a_1 - a_2)}{(b_2 - b_1)}$$

The final equation will always provide a value, except in the case when the gradients (b) of the two lines are equal, in which case the fraction goes to infinity. This is the case of parallel lines, which includes the cases of coincident and collinear lines. Second, the point (p, q) may not lie on the line segments in question, but on the extension of the segment beyond its end points. To test for this, we merely have to test to see if p lies between x_1 and x_2, and if q lies between y_1 and y_2. If the coincidence is close to exact, the intersection is at one of the end points. Function 11.5 implements this algorithm.

Note that Function 11.5 uses two different structures, LINK and POINT. A LINK is a one-dimensional cartographic object connecting two nodes. The POINT structure has been used previously in this chapter. In this function, because the point is to be returned to the calling function, p is a pointer to a POINT structure. This requires the "->" structure member syntax rather than the "." Both LINK and POINT structures were introduced in Chapter 5.

Finally, it should be noted that the intersection algorithm depends upon a value SMALL, which is a tolerance. Setting SMALL to some arbitrary value is not a good idea, and repeated use of the algorithm with different values is suggested as a way to understand the relationship between scale and precision. An alternative tolerance could be one-half of the length of the shortest line segment or the step size of the resolution determined from the data quality report.

Function 11.5 actually uses two different data structures to store the information contained in the line segments. Representation one is with the points of the line stored as a string in the LINK structure. In representation two, the line segments are represented by the constants of the linear equation for the line segment. In our representation, it is still necessary to store the points in the LINK to contain the segment along its line equation.

Saalfeld (1987) in responding to Douglas (1974) listed six ways in which a line can be represented in intersection algorithms. These include that used above, the point-slope form, and the slope-intercept form, plus the two-point form, an alternative two-point form, a linear equation form, and a point-vector form.

Thus a line segment starting at (x_1, y_1) and ending at (x_2, y_2) as in Figure 11.7 can be stored and used by intersection algorithms in any of the following ways:

1. Point-slope form

$$y - y_1 = b(x - x_1)$$

Function 11.5

```
/* intersect: Find the point of intersection between
* two lines, if it exists. Returns 0 for no intersection,
* else 1 and point x,y. kcc 10-88 revised 8-93 */
#include "cart_obj.h"
#define SMALL 0.000001
int intersect(line1, line2, p)
   LINK line1, line2; POINT    *p;
{
int  vertical_case = 0, collinear_case = 0;
float xdif, a1, a2, b1, b2, hi1x, hi2x, hi1y;
float hi2y, lo1x, lo2x, lo1y, lo2y;
xdif = line1.end.x - line1.start.x;
if ((xdif * xdif) < SMALL) vertical_case = 1;
   else { b1 = (line1.end.y - line1.start.y) / xdif;
   a1 = line1.start.y - (b1 * line1.start.x); }
xdif = line2.end.x - line2.start.x;
if ((xdif * xdif) < SMALL) vertical_case += 2;
   else { b2 = (line2.end.y - line2.start.y) / xdif;
   a2 = line2.start.y - (b2 * line2.start.x); }
switch (vertical_case) {
   case 0: /* Neither link is vertical */
      if ((b1 - b2) >= SMALL)
      { p->x = (a2 - a1) / (b1 - b2);
       p->y = a1 + (b1 * p->x); }
      else { /* The two links have equal slopes */
         if ((a1 - a2) >= SMALL)
         /* Collinear links */ collinear_case = 1;
         /* The links are parallel and not collinear */
         else return (0); }
      break;
   case 1: /* First link is vertical */
           p->x = line1.start.x;p->y =a + b2 * p->x;break;
   case 2: /* Second link is vertical */
      p->x = line2.start.x; p->y = a1 + b1 * p->x; break;
   case 3: /* Both lines are vertical */
   if ((line1.start.x - line2.start.x) >= SMALL)return (0);
      else /* Lines are vertical and collinear */
      collinear_case = 1; break;
   default: printf("Error, links do not match case set\n");
   return (-1);
```

Function 11.5 (continued)

```
/* Store x ranges of first link */
hi1x = line1.end.x; lo1x = line1.start.x;
if (line1.end.x < line1.start.x)
    { hi1x = line1.start.x; lo1x = line1.end.x; }
/* Store x ranges of second link */
hi2x = line2.end.x; lo2x = line2.start.x;
if (line2.end.x < line2.start.x) {
      hi2x = line2.start.x; lo2x = line2.end.x; }
/* Store y ranges of first link */
hi1y = line1.end.y; lo1y = line1.start.y;
if (line1.end.y < line1.start.x)
    { hi1y = line1.start.y; lo1y = line1.end.y; }
/* Store y ranges of second link */
hi2y = line2.end.y; lo2y = line2.start.y;
if (line2.end.y < line2.start.y)
    { hi2y = line2.start.y; lo2y = line2.end.y; }
if (collinear_case) { p->y = lo2y; p->x = lo2x; }
/* Test to see if intersection point falls on both links */
if ((p->x <= hi1x) && (p->x >= lo1x) && (p->y <= hi1y) &&
    (p->y >= lo1y) && (p->x <= hi2x) && (p->x >= lo2x) &&
    (p->y <= hi2y) && (p->y >= lo2y))   return (1);
else return (0);
}
```

2. Slope-intercept form

$$y = a + bx$$

3. Two-point form

$$\frac{(y - y_1)}{(x - x_1)} = \frac{(y_2 - y_1)}{(x_2 - x_1)}$$

4. Two-point form (avoids dividing by zero)

$$(y - y_1)(x_2 - x_1) = (x - x_1)(y_2 - y_1)$$

5. Linear equation, using floating point constants a, b, and c.

$$ax + by + c = 0$$

6. Point-vector form, for vector (v_1, v_2) and floating point number r

$$\begin{bmatrix} x & y \end{bmatrix} = \begin{bmatrix} x_1 & y_1 \end{bmatrix} + \begin{bmatrix} rv_1 & rv_2 \end{bmatrix}$$

As in Function 11.5, Saalfeld noted that line segment intersection algorithms mostly use a two-stage approach.

First, a point of intersection is determined, and second the point is tested for inclusion on the two test segments. Computational drawbacks of the algorithms are dealing with parallel or vertical lines, dealing with lines which are almost parallel or vertical, and rounding error. The six representations above all handle the parallel and vertical cases differently and so have computational limitations and costs accordingly.

The rounding and nearly parallel/vertical cases can be solved by using integer coordinates rather than real numbers and by avoiding large numbers. This may mean rounding to meters in UTM, multiplying temporarily by a scaling factor, and truncating large UTM or other coordinates, effectively using a false origin for the computations. Saalfeld pointed out the advantages of the point-vector form, primarily that each value of r corresponds to a unique locations along the line and that r can be made to vary from zero to one.

Distance of a Point to a Line

Yet another critical point-related transformation in automated cartography is establishingthe relationship between a point and higher-dimension objects, such as lines and areas. To establish the distance between a point and a line, it is necessary to determine which link or line segment within the line is closest to the point. This can be done by using the point-to-point distance algorithm above, but may not result in the correct segment (Figure 11.8). A better test is to choose the point-to-segment triangle with the minimum area.

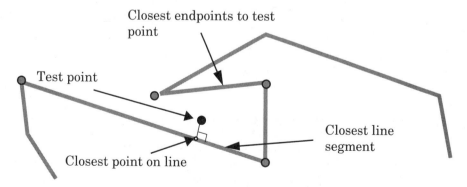

Figure 11.8 Computing the closest point on a line.

Next, the closest links (and there may be more than one, depending on the line) should be tested to determine the closest point along the line to the test point. An easy way to find the intersection point on the line is to mirror the point about the line and then to use the line intersection algorithm above to yield the point (Figure 11.9). Note that the result may be a point that is not on the line segment in question, in which case it should be discarded.

No program for this algorithm is presented here, because the more common cartographic problem is to generate a second line that is at some distance from the first, the so-called line-buffer problem (Schwarz, 1986).

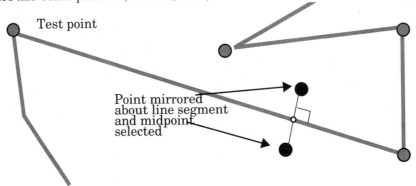

Test point

Point mirrored
about line segment
and midpoint
selected

Figure 11.9 Computing the closest point on a line by mirroring.

11.3.3 Transformations Based on Areas

Area of a Polygon

How do we take a closed line (or G-ring) and calculate area? This is an area to a scalar transformation or simply a measurement. Say that we have a closed polygon with coordinates (x_1, y_1) to (x_n, y_n) and we repeat (x_1, y_1) to close the polygon, let's give the formula and then prove empirically that it works:

$$A = \frac{1}{2} \left| \sum_{i=1}^{npts+1} (x_i y_{i-1}) - (x_{i-1} y_i) \right|$$

Function 11.6 is a computer program to compute the area of a closed polygon stored in our predefined GRING structure (Chapter 7). The returned value will be negative if the polygon is defined counterclockwise and the absolute operator is ignored. Figure 11.8 and Table 11.1 show a simple polygon and the computation of the area by two methods, an intuitive proof and use of the equation. For simplicity, let the products of consecutive x and y values be as follows:

$$A = x_i y_{i-1} \quad \text{and} \quad B = x_{i-1} y_i$$

Table 11.1   Area of A Polygon

Vertex	x	y	A	B	Difference (A–B)
1	1	1	—	—	—
2	1	6	1	6	–5
3	3	6	18	6	12
4	3	5	18	15	3
5	2	5	10	15	–5
6	2	4	10	8	2
7	3	4	12	8	4
8	3	3	12	9	3
9	2	3	6	9	–3
10	2	2	6	4	2
11	3	2	6	4	2
12	3	1	6	3	3
13	1	1	1	3	–2
					16

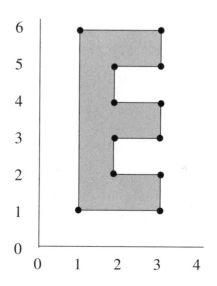

Figure 11.10 Sample polygon for the worked example.

Notice that it is easy to accumulate area as we go, so the computational values can be accumulated while we digitize a polygon. By simple inspection of Figure 11.10 we can see that the polygon consists of a letter E. This letter can be made up from eight unit-square building blocks, five forming a vertical column and three attached to the right in alternate rows.

Clearly, by inspection, the area of the polygon is eight units. From Table 11.1 we can see that the sum of the difference column is 16. If we take half this value, as required in the formula, we get the correct area of eight units. Also note that the value of the area comes out negative if the polygon is specified counterclockwise.

One way we can use this is to include holes within a polygon as rings within the polygon, except specified in the other direction. Then to find the total area, we simply add the area to the area of its enclosed holes to get the actual area of the polygon. Holes within the polygon under the Spatial Data Transfer Standard are considered as rings, with no retraced lines.

Function 11.6

```
/* Function polygon area : Function 11.6
/* Return clockwise area enclosed by RING structure
/* kcc 10-88 revised 8-93 */

#include "cart_obj.h"

float area_of_ring (ring)
GRING ring;
{
  double sigma = 0.0;
  int i;
  for (i=1;i<ring.string.number_of_points;i++)
  {
    sigma +=(ring.string.point[i].x * \
    ring.string.point[i-1].y) - \
    (ring.string.point[i-1].x * ring.string.point[i].y);
    return (0.5 * sigma);
  }
}
```

When the primary data structure is the chain, such as in a topological data structure, the sum of the cross products terms of the segments within each chain can be stored with the chain as a partial solution, to be aggregated when chains are connected together to form a polygon. One problem with this, however, is that the accumulated sums can be very large numbers, especially if the coordinates are values such as UTM coordinates.

Point-in-Polygon Test

The next highest dimension object for which point relations can be established is the area or polygon. There are many algorithms for testing whether or not a point falls within a polygon. Many of the algorithms use area computations, and many others work only for convex polygons or polygons without concavities. These algorithms are virtually useless in analytical cartography, for the shapes of real-world cartographic entities are far from geometric and are almost never entirely convex in their boundaries.

Probably the simplest algorithm for point-in-polygon testing is the Jordan arc theorem. This simply states that a line between a point known to be outside a polygon will cross the polygon boundary an even number of times if the point is outside the polygon and an odd number of times if the point is inside the polygon (Figure 11.11)

This theorem provides solutions in all cases except when the lines either touch a vertex or run parallel to an edge. The parallel problem is often significant, because the outside point can simply be chosen as vertically above or below any given test point, and because many map lines run parallel to the axes.

Function 11.7 works directly for points on the interior area of the polygon. It establishes a point that is outside the polygon by finding the maximum x value for the polygon and multiplying it by 2.

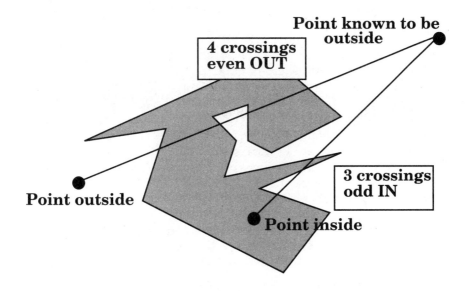

Figure 11.11 The Jordan Arc theorem.

Function 11.7

```
/* Function to perform point in polygon test using
 * Jordan arc theorem     kcc 5-88 revised 8-93
 Function 11.07 */
#include "cart_obj.h"
int point_in_polygon (point, polygon)
POINT point;
RING polygon;
{
POINT pt;
LINE_SEGMENT test_segment, edge;
int i, intersections = 0;
test_segment.point[0].x = point.x;
test_segment.point[0].y = point.y;
/* Establish a point outside the polygon */
test_segment.point[1].y = polygon.string.point[0].y;
test_segment.point[1].x = polygon.string.point[0].x;
for (i=1; i < polygon.string.number_of_points;i++)
  if (polygon.string.point[i].x > test_segment.point[1].x)
    test_segment.point[1].x = polygon.string.point[i].x;
}
test_segment.point[1].x *= 2.0;
/* Now check the link from the test point to the outside
 * point against each of the polygon's edge segments */
    for (i=1; i < polygon.string.number_of_points;i++)
       {
       edge.point[0].x = polygon.string.point[i-1].x;
       edge.point[0].y = polygon.string.point[i-1].y;
       edge.point[1].x = polygon.string.point[i].x;
       edge.point[1].y = polygon.string.point[i].y;
       if (intersect(test_segment, edge, pt))
           intersections++;
}
    /* Return odd/even, 1 means inside, 0 outside */
    return (intersections % 2);
}
```

Other implementations of the algorithm simply choose a point with the same y as the test point but a value of x greater than the largest x value in the ring. This makes the line intersection test somewhat easier, because no intersections have vertical line segments that would have to be treated as a special case in the line intersection routine in Function 11.5.

These algorithms can then deal with the delinquent cases mentioned previously. Sedgewick (1983, p. 317) for example, sorts the points in the ring so that they start with the point with the lowest x value among the points with the lowest y value. Both ends of the test segment are then tested to see if they are on opposite sides of the test segment. Only when this is the case is the intersection counted as the edge of the ring is traversed.

Thiessen Polygons

The final transformation between a point and a higher-dimension object to be considered in this section is a point-to-polygon transformation. The earliest computer cartographic program to perform this transformation was the SYMAP package, which called the polygons created *proximal regions*. These regions are called *Thiessen polygons* in geography after climatologist A. H. Thiessen. In computer science and mathematics the term *Voronoi diagram* is used. Thiessen devised these regions to assist in the interpolation of climatic data from unevenly distributed weather stations.

The method is used wherever data have been collected at points and the cartographer desires to use area-based analytical techniques or symbolization methods. As such, this is a space-partitioning transformation, because it divides the regions surrounding a set of points exclusively into a set of polygons.

The algorithms for finding the Thiessen polygons are too numerous and too lengthy to be included here; the reader is referred to a recent review by Tsai (1993). One noteworthy side effect of computing these areas is that the Delaunay triangulation is also performed because it is the dual of the Voronoi diagram. This transformation partitions the space using the points as nodes in the network and is often used to produce a TIN (triangulated irregular network) structure. Other applications are to make choropleth type maps of point data and to assist in interpolation over space.

11.4 AFFINE TRANSFORMATIONS

Map projections are the only coordinate transformations covered so far. It is possible, if we assume plane geometry, to provide a set of transformations for conversion between any two bounded planar Euclidean coordinate spaces. These transformations, known as *affine* transformations, consist of three distinct transformations applied in sequence.

The sequence is, first, a *translation,* in which the coordinate system origin is moved to a new location; second, a *rotation,* in which the axes of the coordinate system are rotated to match the new system; and finally, a *scaling,* in which the numbers along the axes are scaled to represent the new space scale. It is possible to accumulate the effects of these three transformations into a single step, which can then be applied on an object-by-object basis.

First, a transformation to move the origin can be stated as

$$\begin{bmatrix} x & y & 1 \end{bmatrix} \begin{bmatrix} 1 & 0 & 0 \\ 0 & 1 & 0 \\ -x_0 & -y_0 & 1 \end{bmatrix} = \begin{bmatrix} x - x_0 , y - y_0 , 1 \end{bmatrix}$$

Note the similarity between the translation and the movement of the origin for a map projection. Map projections usually start by subtracting from the longitude the longitude at which the projection is to be centered, equivalent to rotating the earth to a new center of view.

The next stage is a rotation of the axes about an angle θ

$$\begin{bmatrix} x - x_0 & y - y_0 & 1 \end{bmatrix} \begin{bmatrix} \cos\theta & \sin\theta & 0 \\ -\sin\theta & \cos\theta & 0 \\ 0 & 0 & 1 \end{bmatrix} =$$

$$\begin{bmatrix} \cos\theta\,(x - x_0) - \sin\theta\,(y - y_0) & \sin\theta\,(x - x_0) - \cos\theta\,(y - y_0) & 1 \end{bmatrix}$$

Third, the point is subject to a scaling of the axes:

$$\begin{bmatrix} \cos\theta\,(x - x_0) - \sin\theta\,(y - y_0) & \sin\theta\,(x - x_0) - \cos\theta\,(y - y_0) & 1 \end{bmatrix} \begin{bmatrix} S_x & 0 & 0 \\ 0 & S_y & 0 \\ 0 & 0 & 1 \end{bmatrix} =$$

$$\begin{bmatrix} S_x[\cos\theta\,(x - x_0) - \sin\theta\,(y - y_0)] & S_y[\sin\theta\,(x - x_0) - \cos\theta\,(y - y_0)] & 1 \end{bmatrix}$$

Note that the scaling also looks much like the scale transform in a map projection. This is precisely what the affine transformation above is doing. The sequence of these transformations is translation, rotation, and scaling. Often order is important; in other words, if we use some other sequence, we will not end up with the same final location. That has implications for inverting this transformation. Because the transformation is multistep and we have to perform the steps in the correct sequence, we also have to undo themin the correct sequence.

We can multiply these 3 by 3 matrices by each other and produce a single transformation which summarizes the entire set of transformations:

$$\begin{bmatrix} x & y & 1 \end{bmatrix} TRS = \begin{bmatrix} x' & y' & 1 \end{bmatrix}$$

In nonmatrix algebra, this yields:

$$x' = S_x[\cos\theta\,(x - x_0) - \sin\theta\,(y - y_0)]$$
$$y' = S_y[\sin\theta\,(x - x_0) + \cos\theta\,(y - y_0)]$$

Figure 11.12 shows these three transformations graphically.

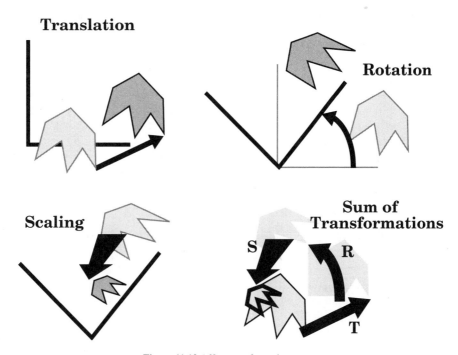

Figure 11.12 Affine transformations.

Note that the x and y of the origin are the x and y of the new origin in the old coordinate system. Consider the following example: A point in the old coordinate system is at (2, 2). What is the coordinate of the point in a system that has a new origin at point (0, 2) in the old coordinate system, a 30-degree clockwise rotation of the two sets of axes, and new axes that have twice the spacing? Using the formulas, we find that

$$x' = 2\left[\cos 30\,(2-0) - \sin 30\,(2-2)\right]$$
$$y' = 2\left[\sin 30\,(2-0) + \cos 30\,(2-2)\right]$$

This reduces to

$$x' = 2\left[0.866\,(2) - 0.5\,(0)\right] = 3.464$$
$$y' = 2\left[0.5\,(2) + 0.866\,(0)\right] = 2.000$$

Typically, these expressions are computed during the setup and registration phase of digitizing, when the affine constants are computed once and stored from a set of control points using the least squares solution given below. Then, successive points can be transformed as they are collected, or as a single step after data collection.

11.5 STATISTICAL SPACE TRANSFORMATIONS

11.5.1 Rubber Sheeting

When data have been captured from a flat map in cartesian coordinates, the affine transformations can be used for transforming data coordinates. In many cases, however, including those in which direct measurement of a cartographic entity has taken place, the precise map geometry is undetermined. This is especially true with remotely sensed images and air photos, when the viewing geometry is unknown. In both cases, the image is distorted for a number of reasons, including the flight geometry, the earth's curvature, the portion of the earth covered, the view angle, the topography, the altitude of the observation platform, the characteristics of the lens, and a host of other reasons. Also, the atmosphere itself is far from uniform, making its own "lens" effects, depending on pressure, moisture content, time of day, solar heating, and so forth.

Even under the best of circumstances, aircraft and even satellites pitch, yaw, and even "wobble." As a result, the earth's surface as seen in an image is an approximate map. In this map the actual cartographic objects and features can be plainly seen, but the precise relationships between scale, area, shape, and direction are imprecise.

An alternative way to state the relationship between such a distorted image and an accurate cartographic map base is to state that the precise map projection transformation is unknown. Instead, we have numerous actual examples of the transformation having been applied. To describe the projection characteristics fully, which is necessary if we wish to invert its effects and convert the image back to actual map space, we have to study the projection empirically.

One analogy is to imagine the map printed on a rubber sheet. When the air photo or satellite image arrives, we should be able to recognize precisely the locations of a key set of places on the image. These locations are the *ground control points*. A ground control point is a physical feature in a scene that is clearly detectable and whose precise coordinates in a coordinate system are known. For some purposes, merely the location of ground control points is enough, but for others we should know the location and elevation quite precisely and with a high degree of accuracy.

Bernstein (1983) noted that typical ground control points are airports, highway intersections, land-water interfaces, and geological and field patterns. Campbell (1987) also noted that the ground control points should ideally be a single pixel in size on the image and should be easily identifiable against their background. They should be dispersed throughout the image, particularly at the edges, not all concentrated in a single region.

To follow the analogy, the ground control points can be seen as holes through which the rubber sheet map is pinned to the correct map and stretched to fit correctly, that is, the pins are at their correct locations on both the map and the image. Given this, we can measure the "stresses" in the rubber sheet to figure out how to stretch it into the map's geometry. We then apply these "stresses" to the image, and as a result the entire image is stretched into the map space. Strictly, the locations of the control points themselves should not be modified by the algorithm, although this is not the case with the function used below (Figure 11.13). Also, the model of error assumed is that the error is uniformly random over the area in question. This is rarely the case when data have come from scanners, instruments, or field measurements.

Mathematically, we have sets of points that match in the two spaces. These points have n locations (x, y) on the map and (u, v) in the image.

$$\begin{vmatrix} x \\ y \end{vmatrix} = T \begin{vmatrix} u \\ v \end{vmatrix}$$

The trick is to use the n multiple cases of (x, y) and (u,v), for which we have a set of

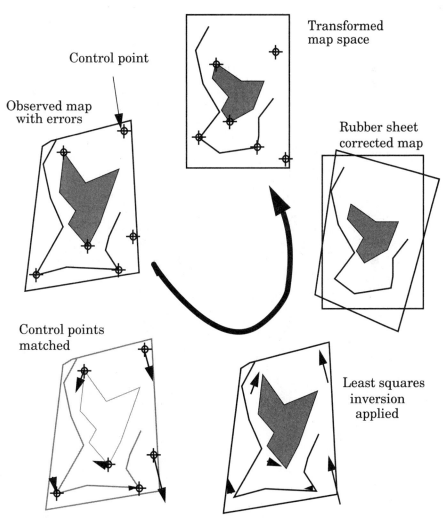

Figure 11.13 Rubber sheet transformations by least squares affine.

ground control points, to estimate T. The process of estimation proceeds by using what Sprinsky (1987) called the six-parameter affine, that is,

$$u = \beta_0 x + \beta_1 y + \beta_2 \qquad\qquad v = \beta_3 x + \beta_4 y + \beta_5$$

The β coefficients can be computed by a sort of least squares fit. If we let

$$p = \frac{\sum_{i=1}^{n} [(x_i - \bar{x})(y_i - \bar{y})]}{\sum_{i=1}^{n} (y_i - \bar{y})^2}$$

and

$$q = \frac{\sum_{i=1}^{n} [(x_i - \bar{x})(y_i - \bar{y})]}{\sum_{i=1}^{n} (x_i - \bar{x})^2}$$

then the following sets of equations allow the solution of the six unknown β values from n control points.

$$\beta_1 = \frac{\sum_{i=1}^{n} [(y_i - \bar{y})(u_i - x_i)] - q \sum_{i=1}^{n} [(x_i - \bar{x})(u_i - x_i)]}{\sum_{i=1}^{n} (y_i - \bar{y})^2 - q \sum_{i=1}^{n} [(x_i - \bar{x})(y_i - \bar{y})]}$$

$$\beta_3 = \frac{\sum_{i=1}^{n} [(x_i - \bar{x})(v_i - y_i)] - p \sum_{i=1}^{n} [(y_i - \bar{y})(v_i - y_i)]}{\sum_{i=1}^{n} (x_i - \bar{x})^2 - p \sum_{i=1}^{n} [(y_i - \bar{y})(x_i - \bar{x})]}$$

$$\beta_4 = 1 - p\beta_3 + \frac{\sum_{i=1}^{n} [(y_i - \bar{y})(v_i - y_i)]}{\sum_{i=1}^{n} (y_i - \bar{y})^2}$$

$$\beta_0 = 1 - q\beta_1 + \dfrac{\displaystyle\sum_{i=1}^{n} [(x_i - \bar{x})(u_i - x_i)]}{\displaystyle\sum_{i=1}^{n} (x_i - \bar{x})^2}$$

$$\beta_2 = \frac{1}{n}\sum_{i=1}^{n}(u_i - x_i) + \bar{x} - \beta_0\bar{x} - \beta_1\bar{y}$$

$$\beta_5 = \frac{1}{n}\sum_{i=1}^{n}(v_i - y_i) + \bar{y} - \beta_4\bar{y} - \beta_3\bar{x}$$

The inverse transformation represented by the inversion of these formulas can then be used to allow the image coordinates as input and to provide map coordinates as output:

$$x = \alpha_0 u + \alpha_1 v + \alpha_2 \qquad y = \alpha_3 u + \alpha_4 v + \alpha_5$$

where, for convenience, we let

$$k = \beta_4\beta_0 - \beta_3\beta_1$$

Then

$$\alpha_0 = \frac{\beta_4}{k} \qquad \alpha_1 = -\frac{\beta_1}{k} \qquad \alpha_2 = \frac{\beta_1\beta_5 - \beta_2\beta_4}{k}$$

$$\alpha_3 = -\frac{\beta_3}{k} \qquad \alpha_4 = \frac{\beta_0}{k} \qquad \alpha_5 = \frac{\beta_3\beta_2 - \beta_5\beta_0}{k}$$

This solution is also frequently used in surveying to adjust errors in measurement over a set of readings, and is frequently developed as a matrix solution (Moffitt and Bouchard, 1982). Several of the terms in the solution above are clearly determinants and transposes. A computer program to illustrate the solution is included on the companion disk. Once more, we have an example of a cartographic transformation. In this case, however, we have had to deduce the exact nature of the transformation from its results, and have inverted the transformation empirically in a "best-fit" solution.

At any point is space, it is possible to compute a vector displacement with an x and a y component from the inverse transformation. This vector can be used to compute the root mean squared error (RMS) at any point in the system. RMS error computation is often averaged over many points, so that an overall measure of error can be computed. Also, the RMS error at control points can be computed and used to eliminate from the computation those control points that contribute most to the RMS error. Thus the six-parameter affine transformation allows both practical transformation into a known geometry and a means of analyzing the error in doing so.

When there is little or no control over the measurement transformation (as in many air photo applications) or when the transformation is very complex, the statistical affine is the only approach available. Finally, it should be noted that in remote sensing the final result is a grid of points that are not regular. The points are therefore immediately resampled back into a regular grid that is orthogonal to the map coordinate system and has some finite pixel size. Methods for performing this second step transformation are discussed in Bernstein (1983).

11.5.2 Cartograms

Cartograms, also known as value-by-area maps and varivalent projections (Tobler, 1986), are maps drawn so that the areas of their internal enumeration units are proportional to the values they represent (Dent, 1985, p. 326). Generally, they are of two types. First, *non-contiguous* cartograms simply shrink the scale factor of an areal unit in isolation from all others and present the results as a diagram in which geographic space is interrupted. As such, these maps could be regarded as a form of proportional point symbol mapping. Second, *contiguous* cartograms maintain the continuity of geographic space, and as such are really a variation upon map projections. Like all map projections, direction, distance, shape, area, and the graticule are distorted in combination on these maps. Their advantage is that the map base can be distorted to make the mapping of a thematic data item more appropriate, removing the underlying differences in population density when showing per capita income, for example.

Tobler (1986) advocated the following method for the computation of "pseudo-cartograms." First, the data to be used to "reshape" the map should be converted to the structure of a uniform grid, with equal latitude-longitude spacing north-south and east-west. If this grid is an m by n array, Z, then it can be written as the sum of k arrays, where k is the lesser of m and n. The arrays can then be written as a set of products of each of two k by 1 vectors, U and V :

$$Z = U_1 V_1 + U_2 V_2 + ... + U_k V_k$$

where U_k is the kth eigenvector of the matrix Z multiplied by its transpose, Z^T and

$$V_k = \frac{Z^T U_k}{U^T U_k}$$

The vectors can then be sorted by the amount of variance accounted for by the model, and the first picked to yield the transforms. The U vectors are now functions only of the row index or y, and the V vectors are functions of the column index or x. These values are then projected using a standard equal-area map projection:

$$y' \ = \ Rs U_1 \cos y \qquad x' \ = \ Rs V_1$$

Tobler has published extensively on these transformations, including a set of computer programs (Tobler, 1974).

11.6 SYMBOLIZATION TRANSFORMATIONS

As far as producing a map is concerned, the final transformation is the symbolization transformation. In this transformation, the cartographic objects are subjected to all the necessary transformations to use a specific map type, and the map itself is generated. Although the actual look of the map is determined largely by the type of data, the fundamental properties of the data, the type of map, and the design of the map, there are some common or generic symbolization transformations. Among these is the viewing transformation, in which the geometry of the map is established. Also important is the actual plotting of the graphic elements, how they are symbolized, and how the text is added. This section contains information about these final, often critical, symbolization transformations according to the GKS standard.

11.6.1 The Normalization or Viewing Transformation

A point-to-point transformation of great importance in analytical and computer cartography is the transformation between the *world* coordinates, given in the coordinate system required by an application, and the coordinates of a particular display device, such as the screen of a graphics terminal. Typical *world* coordinate systems are either arbitrary (map millimeters, for example), or standardized (such as UTM coordinates). The user should note that world coordinates as defined by the GKS standard are not the same as geographic (latitude and longitude) coordinates but can be any coordinate system into which the objects have been geocoded. An orthogonal planar coordinate system is assumed. The display device also uses arbitrary coordinates known as *device* coordinates, usually the number of pixels that are addressable in the east-west and north-south directions. Because each particular display device has a different number of addressable pixels, it is desirable to have an intermediate coordinate system with a standard size into which we can transform the map.

The intermediate coordinates are known as *normalized device* coordinates. In the language of the Graphical Kernel System, the transformation from the world to the normalized device coordinates is known as the *normalization* transformation, while the transformation to device or screen coordinates is the *workstation* transformation. These transformations are summarized in Figure 11.14.

The normalization transformation is specified by defining the limits of an area in the world coordinate system known as the *world window*. This rectangle or square is mapped onto a specified part of the normalized device coordinate space called the *world viewport*.

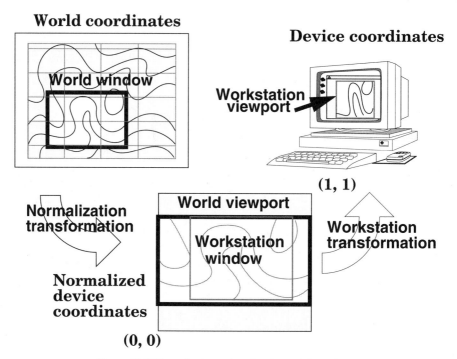

Figure 11.14 Normalization and workstation transformations.

Window and viewport limits are assumed to be parallel to the coordinate axes, so any rotations must be handled by the user, using the rotation algorithms given previously. Normalized device coordinates range from (0.0, 0.0) at the lower left corner to (1.0, 1.0) at the upper right corner. Similarly, we can control how much of the normalized device coordinate space is visible by establishing a workstation transformation. This is done by specifying a workstation window in normalized device coordinates. The actual transformations, the mathematics of which can be derived from the transformational equations given in the previous section on affine transformations, are contained within the Graphical Kernel System functions `set_window` and `set_viewport`.

A small interactive program to establish the transformation is given as Function 11.8. In this example, the goal is to center the map onto the largest possible piece of the device space while maintaining the correct scale relationships between x and y. This is done with the help of the GKS inquiry function `inquire_display_space_size`. This function returns the device coordinate units, the size of the surface of the device, and the number of pixels the device is capable of displaying. Note that your particular implementation of GKS will vary slightly in how C-binding GKS functions are named. SunGKS, for example, uses `gsetwindow`, `gsetviewport` and so forth. In a windowing system or graphical user interface, the dimensions of the device become those of the active window.

Function 11.8

```
/* Establish the normalization transformation for an active
 * display device, using the largest available centered
 * rectangle with appropriate scale.
 * Function 11.08 kcc    5-88 Revised 8-93 */
#include "gks.h"
void normalize()
{
    float llx, lly, urx, ury, xmin, xmax, ymin, ymax;
    float units, xpix, ypix, width, height, aspect_ratio,
        map_size;
    /* Establish the size of the display */
    ginquire_display_surface_size (&units, &xmax, &ymax,
        &xpix, &ypix);
    printf(" Enter the world window\n");
    printf(" Enter (xmin, ymin) :");
    scanf ("%f%f%*1c", &llx, &lly);
    printf(" Enter (xmax, ymax) :");
    scanf ("%f%f%*1c", &urx, &ury);
    /* Establish the aspect ratio of the world window */
    width = urx - llx; height = ury - lly;
    aspect_ratio = height / width;
    /* Apply the same aspect ratio to the display device */
    if (aspect_ratio < 1.0) { /* Landscape image */
        map_size = xmax * aspect_ratio;
        ymin = (ymax - map_size) / (2.0 * xmax);
        ymax = (ymin + map_size) / xmax; xmax = 1.0;
    } else { /* Portrait Image */
        map_size = ymax / aspect_ratio;
        xmin = (xmax - map_size) /( 2.0 * ymax);
        xmax = (xmin + map_size) / ymax; ymax = 1.0 }
    /* Finally set up the normalization transformation */
    set_window ( llx, urx, lly, ury);
    set_viewport ( xmin, xmax, ymin, ymax);
    return;
}
```

11.6.2 Primitives and Attributes

Under GKS, one of the basic tasks of the system is to generate *pictures,* the standard name for the final rendered or symbolized graphic. In the case of computer cartography, the pictures contain symbolizations of cartographic objects, that is, maps. In the terminology of the Spatial Data Transfer standard, the real-world phenomenon to be represented on a map is called a cartographic entity. The digital representation, usually as a data structure, of the entity is as a cartographic object. In the final cartographic transformation, the *symbolization* transformation, the cartographic object is depicted as a cartographic element, that is, a distinct part of the graphic of a map. As such, the cartographic map element is a picture under GKS terminology and therefore consists of a collection of *primitives* with their associated *attributes.*

GKS defines and makes available to the programmer six primitives for digital drawing. These consist of one line primitive, one point primitive, one text primitive, two raster primitives, and one general-purpose primitive (Figure 11.15). The final cartographic transformation is to convert the cartographic data into groups of these primitives and to set their attributes. The line primitive is called the *polyline.* In symbolizing a polyline, GKS generates a set of straight line segments connecting a given point sequence. The point primitive is the *polymarker.* To symbolize a polymarker, GKS generates a symbol of a given type at a specified location. Symbols can be points, circles, squares, and so forth. The text primitive in GKS is *text.* Text causes GKS to generate a text character string at a specified location.

The two raster primitives under GKS are *fill area* and *cell array.* Fill area is used to generate polygons and to control the symbolization of their interiors. Cell array generates an array or grid of rectangular cells, each with individual colors. The cell array is an abstraction of the array of pixels on a raster type device. Finally, the *generalized drawing primitive* is designed to make use of special hardware characteristics of a specific workstation. Some workstations can generate circles, ellipses, spline curves,and so forth. These primitives have the workstation geometric transformations applied to them, but actual plotting is left up to the workstation.

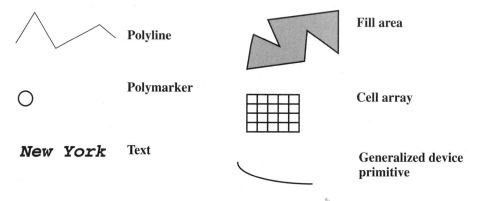

Figure 11.15 Primitives in the GKS standard.

Attributes are a function of the primitive to which they apply. Generally, polylines have the attributes of *line width, line type,* and *line color.* Polymarkers have the attributes of *marker size scale factor, marker type,* and *color.* Text has by far the most attributes, including *character height, character up vector, character expansion factor, text path, character spacing, text font, text precision, color,* and *text alignment.* Fill areas have the attributes of *pattern size, pattern reference point, pattern array, interior style, hatch style,* and *color.* The cell array has only *color,* while the attributes of the generalized drawing primitive are dependent on its type. Another important attribute is the *pick identifier,* which controls the way the user may interact with the particular primitive when it is displayed on the workstation, allowing objects to be interactively selected by the user.

Among the large variety of possible settings, line types may be solid, dashed, or dashed-dotted; text path may be left, right, up, or down; character alignment may be left-aligned, centered, or right-aligned to top, normal (middle), or bottom; and interior styles may be hollow, solid, pattern, or hatch. In addition, the attributes may be individual, in which case they apply to all the primitives that follow until they are changed, or bundled. If bundled, the attributes may be collected together as a group and applied collectively to one or more primitives. Such a collection of primitives is stored for an application such as a mapping program in a *bundle table,* a set of linked attribute settings for different but related primitives. *Indexes* named for each primitive allow separate management of these bundle tables.

It should be noted that GKS implementations usually come with library files, which predefine certain attribute parameters so that more intuitive names can be used. These definitions are often in a header file, such as `gksdef.h`, but the actual name depends on the GKS implementation. The pre-processor names, in keeping with the C standard, are in uppercase. So, for example, the first eight colors, which require calls with integers 1 through 8, are instead `#defined` to RED, GREEN, BLUE, MAGENTA, CYAN, YELLOW, BLACK, and WHITE. Similarly, the GKS implementations come with a documented set of supported output devices, such as X-windows terminals, PostScript output and the Computer Graphics Metafile. These values are all fully determined by the attribute `workstation_type`.

11.6.3 Drawing Cartographic Objects

The very final stage in the cartographic transformation from data to map is to actually implement a primitive, after having set its attributes, either individually or in a bundle. Before the primitives can be accessed, a workstation must be open and the sets of geometric space transformations to be applied must be active. Assuming that this is the case and that the computer cartographer has loaded data into the data structures we have outlined in earlier chapters, all that remains is to point to the GKS functions that actually do the display. First, we will cover the actual calls themselves, and following each function will be the relevant attributes that can be changed to manipulate the workstation display.

1. Polyline

To display a polyline, as stored in the previously defined STRING data structure, the

GKS call is

```
polyline(String.number_of_points,&String.point[0].x,
        &String.point[0].y);
```

Attributes are

```
set_linewidth_scale_factor (scale);
```
set_linetype (type);

2. Polymarker

A single polymarker is defined and stored in the previously defined POINT structure. Of course, these are normally handled as arrays of polymarkers. The GKS drawing call is

```
polymarker (1, Point.x, Point.y);
```

Attributes are

```
set_marker_size_scale_factor (scale);
set_marker_type (type);
set_marker_color_index (color);
```

3. Text

Text has not been treated as having a cartographic data structure in this book. In its simplest form, text consists of a string of C type `char`. The GKS drawing call is

```
text (text_string);
```

Attributes are

```
set_text_alignment (horizontal, vertical);
set_text_font_and_precision
            (font_idex,precision_index);
set_text_path (path_direction);
set_text_color_index (color);
```

4. Fill area

To fill a polygon, the cartographer should use the previously defined RING structure, into which has been loaded data. The GKS command to plot the polygon is

```
fill_area (Ring.number_of_points, &Ring.point[0].x,
          &Ring.point[0].y);
```

Attributes are

```
set_fillarea_color_index (index);
set_fillarea_interior_style (style);
set_pattern_reference_point (x, y);
```

```
set_pattern_size (size_x, size_y);
```

To fill a complex polygon, the polygon boundary should be filled first, followed by the holes.

5. Cell array
To depict an array, such as a digital elevation model or a hill-shaded grid on a workstation, the cartographer should use the previously defined GRID structure, into which has been loaded data. The GKS command to plot the entire grid cell map is

```
grid_cell (x_p, y_p, x_q, y_q, no_cols, no_rows, col_start,
           row_start, dx, dy, &Grid.z[0][0]);
```

The only attribute is

```
set_color_index (index);
```

which should be called color by color to set the color table.

6. Generalized drawing primitive (GDP)
The depiction of a GDP is specific to a particular workstation, and is rarely used under cartographic applications unless drawing times become too slow.

11.7 SUMMARY

In this chapter we have covered numerous examples of cartographic transformations. We have worked through specific examples of how a transformational point of view allows the organization of the body of material that makes up computer cartography into its constituent parts within analytical cartography. Two major types of transformations have been presented. First, we have seen how cartographic objects can be transformed by dimension. As part of these dimensional transformations, the case of point-to-point transformations was used to show how the locational attributes of cartographic data can be transformed to produce map projections and cartograms. Second, we have examined the symbolization transformation, in which the cartographic objects gain actual instances as GKS pictures, collections of graphics primitives with their associated attributes.

Transformations of the first type noted above are fully within the domain of analytical cartography. The symbolization transformation is the very essence of computer cartography, and we have made the natural step from discussion of algorithms directly to computer standards and functions. The symbolization transformation and the programming theme form Part IV of this book, Producing the Map. In the remainder of Part III, Chapters 12 and 13, we will look specifically at line and area transformations, at data structure transformations, and finally at the transformations possible with volumetric data.

11.8 REFERENCES

Anderson, P. S. (1994). *Guide for Users of MicroCAM*, February edition, Project Micro-CAM, Department of Geography-Geology, Illinois State University, Normal, IL.

Bernstein, R., ed. (1983). "Image Geometry and Rectification" in *Manual of Remote Sensing. Volume I: Theory, Instruments and Techniques*, edited by D. S. Simonett, American Society of Photogrammetry, Falls Church, VA: Sheridan Press, pp. 873–922.

Campbell, J. B. (1987). *Introduction to Remote Sensing.* New York: Guilford Press.

Dent, B. D. (1985). *Principles of Thematic Map Design.* Reading, MA: Addison-Wesley.

Douglas, D. H. (1974). "It Makes Me So CROSS." Harvard University Laboratory for Computer Graphics and Spatial Analysis, internal memorandum. Reprinted in D. Marble, H. Calkins, and D. Peuquet (1984) *Basic Readings in Geographic Information Systems.* New York: SPAD Systems, Ltd.

Microcomputer Specialty Group. Association of American Geographers. MicroCAM Software. Contact Dr. Robert P. Sechrist, Dept. of Geography, Indiana University of Pennsylvania, Indiana, PA 15705.

Moffitt, F. H., and Bouchard, H. (1982). *Surveying.* 7th ed. New York: Harper & Row.

National Oceanic and Atmospheric Administration (1988). *The General Cartographic Transformation Package.* National Ocean Service, Charting and Geodetic Service, National Geodetic Survey, Rockville, MD.

Saalfeld, A. (1987). "It Doesn't Make Me Nearly as CROSS. Some Advantages of the Point-Vector Representation of Line Segments in Automated Cartography." *International Journal of Geographical Information Systems,* vol. 1, no. 4, pp. 379 - 386.

Schwarz, C. R. (1986). "Algorithms for Constructing Lines Separated by a Fixed Distance." Proceedings of the Second International Symposium on Spatial Data Handling, Seattle.

Sedgewick, R. (1983). *Algorithms.* Reading, MA: Addison-Wesley.

Snyder, J. P. (1983). *Map Projections Used by the U.S. Geological Survey,* 2d ed. Geological Survey Bulletin 1532. Washington, DC: U.S. Government Printing Office.

Snyder, J. P. (1987*). Map Projections—A Working Manual.* Geological Survey Bulletin P-1395. Washington, DC: U.S. Government Printing Office.

Snyder, J. P., and Voxland, P. M. (1989*). An Album of Map Projections.* Geological Survey Professional Paper 1453. Washington, DC: U.S. Government Printing Office.

Sprinsky, W. H. (1987). "Transformation of Positional Geographic Data from Paper-based Map Products." *American Cartographer,* vol. 14, no. 4, pp. 359–366.

Tobler, W. R. (1974*). Cartogram Programs.* Department of Geography, University of Michigan, Ann Arbor.

Tobler, W. R. (1979). "A Transformational View of Cartography." *American Cartographer,* vol. 6, no. 2, pp. 101–106.

Tobler, W. R. (1986). "Pseudo-cartograms." *American Cartographer,* vol. 13, no. 1, pp. 43–50.

Tsai, V. J. D. (1993). "Delaunay Triangulations in TIN Creation: An Overview and a Linear Time Algorithm." *International Journal of Geographical Information Systems,* vol. 7, no. 6, pp. 501–524.

12

Data Structure Transformations

12.1 WHY TRANSFORM BETWEEN STRUCTURES?

In virtually all mapping applications it becomes necessary to convert from one carto-graphic data structure to another. The ability to perform these object-to-object transfor-mations often is the single most critical determinant of a mapping system's flexibility. Why is this the case? A number of reasons dictate that the mapping process involve data structure conversions, and these are related to data input and geocoding, data storage and representation, the suitability of particular structures for different analytical and model-ing demands, and finally the demands of a particular symbolization transformation.

Data input has already been noted as a primary determinant of data structure. Partic-ular data capture devices, such as scanners and semiautomatic digitizers, generate data in a specific form. Scanners, for example, generate grids, and semiautomated digitizers pro-duce strings of (*x, y*) coordinates. Geocoding stamps a particular coordinate system, res-olution, and map projection on the data. In virtually every case, the cartographer will find that the available digital cartographic data are in the wrong structure for the required type of cartographic object, are on the wrong map projection, or have the wrong resolution for mapping or that the map rectangle needs to be rotated, scaled, or translated to produce the map.

Analytical and computer cartography, unlike many other disciplines, have available large reserves of common, generically digitized cartographic data. As a result, cartogra-phers must change data structures simply to move the data into the correct geographic ex-tent and from an input format into the format required by a particular piece of mapping software.

In Chapter 5, consideration was given to the different mechanisms by which digital cartographic data can be stored within a computer's memory and to how different means of data representation can save storage or improve data accessibility. Because one of the distinguishing characteristics of cartographic and geographic data is sheer volume, the conversion of data between structures, or between representations of a single structure, can save considerable storage space and processing time. The time and space limitations become more apparent as cartographic software moves from larger to smaller computers.

Although the amount of RAM and disk storage available to microcomputers has increased and mass storage technology is making significant breakthroughs in volume capabilities, processing the millions of data points necessary for high-accuracy cartography really strains the microcomputer. Precision or scale are usually sacrificed, with cartographically unacceptable results. Transforming data between cartographic data structures means that the application can optimize storage and processing time as is appropriate. Often the best data structure for a map depends on the demands placed on the data for analytical or display purposes. Good cartographic software does not force data into a single structure, but retains the option to flip between structures on demand.

Analysis and modeling also require different data structures. As an example, the skeleton or medial axis transform of a polygon can be performed in both grid or polygon entity-by-entity mode. In entity-by-entity mode, the locus of the largest enclosed circles must be traced through the polygon. In grid mode, the polygons are simply eroded away step by step from the edges while connectivity is maintained, until the medial axis is formed. Each operation is fairly fast. Inverting the transformation in grid mode is simple. Rebuilding the polygon from the medial axis, however, is difficult in entity-by-entity mode. The grid data structure is suitable for modeling and analytical operations such as Fourier and principal component analysis, filtering, and edge detection.

The TIN structure is good for modeling overland and stream hydrology and for simulating erosion. Entity-by-entity definitions are good for high-precision output, with multiple-weight lines and elaborate shading. Continuous patterns are more suited to the grid. The relative advantages and disadvantages of the various structures are many and even depend on the implementation characteristics, such as language, computer, and operating system.

For symbolization, actually producing the map, again certain structures meet different sets of cartographic demands, which implies that different types of map, different map scales, and so forth, all have their different optimal cartographic data structure. Often, the characteristics of the output device determine the best data structure. For example, raster devices favor grid and quad tree structures, while plotters, laser-jet printers, and automatic scribers require data in line or polygon entity-by-entity format.

The type of representation is also important. Choropleth maps can be produced in any structure, but hill-shading and perspective views are best using grids or TINs. As far as symbolization is concerned, nowhere is the influence of data structure more obvious than in map text labeling. Vector displays usually use the Hershey fonts and are capable of some quite attractive precision lettering.

Raster devices use bit-mapped graphics and as such produce blocky text (which looks worse with enlargement). In addition, the bit-mapped graphics are restrained by their angle on the screen. Just as many systems now support both raster and vector, the now common PostScript graphics merge vector draw commands with the raster capabilities of laser-jet printers. Some paint programs support bit-mapped (raster) or "object" (vector) modes within a single structure. It is with the text, again, that the differences are most apparent, especially with enlargement.

Clearly, there are many reasons to transform digital cartographic data between data structures. As such, we move from one type of cartographic object to another. Changing data structure can result in a loss of spatial information. These data structure transformations may not be fully invertible. Information loss can come as the result of changing scale, termed here *resampling*, as a direct consequence of changing data structure, or as a consequence of the data structure transformational process or the algorithm itself. The study of data structure transformations, and of course their perfection as a consequence of their study, is an important goal for analytical cartography. In this chapter we will first consider scale or resampling transformations, then the specific transformation between vector and raster data structures. A classification of cartographic data structure transformations and their inverses will then be considered. To conclude, we will discuss the role of error in data structure transformation.

12.2 GENERALIZATION TRANSFORMATIONS

Generalization transformations are those in which the scale or equivalent scale of cartographic objects is changed. Cartographic generalization is usually undertaken for one of two reasons: first, so that a set of cartographic objects can be symbolized or used analytically at a scale different (usually smaller) from that at which geocoding took place; and second, to generalize a map either for clarity of symbolization or for the reduction of the data set size. It is important to realize that this transformation applies only to the map data, rather than to the attribute data involved.

At several stages in the discussion of cartographic data structures it has been necessary to consider separately point, line, area, and volume data. Resampling transformations are within their own dimensional type, such as point to point, rather than between types, such as area to point.

12.2.1 Point-to-Point Transformations

A point-to-point transformation involves selecting one point to represent multiple points. One such point is the centroid, discussed in Chapter 11. The centroid, perhaps with a scatter parameter, is a summary of the locational characteristics of a point distribution. A primary use for the centroid is in the conversion of irregularly spaced data, or data collected throughout a set of regions, into a continuous representation such as a grid. Population density data, for example, are often computed for census tracts in a city and converted to a grid by interpolation from point centroids selected in some manner. The point-to-grid transformation will be covered in detail in Chapter 13.

Many grid-to-grid transformations are, in fact, point to point, as are changes between map projections and coordinate systems. A transformational process that uses as input a global data set of latitudes, longitudes, and elevations, usually organized at regular intervals of degrees, minutes, or seconds, does not produce a regular grid after a map projection transformation. After the transformation into a map projection, the data must be resampled into a grid based on the axes of the new coordinate system. Because point-to-point transformations are so common, they are usually thought of as part of the geocoding process. They are, however, important examples of cartographic transformations.

12.2.2 Line-to-Line Transformations

Line-to-line transformations have received considerable attention. A general statement of the problem would be to take a cartographic line, as represented as an object in a particular data structure, and to reduce the number of elements required to store the line in such a way that the line symbolization produced carries the spatial properties of the line to the map reader. Line character is important to preserve, yet difficult to define, and involves a complex set of related approaches to generalization (Buttenfield, 1985).

Function 12.1 performs the *n*th point retention generalization on a string in the STRING data structure from Chapter 5. The function read_strings() was introduced in Chapter 7. These functions are used with the main program in Function 12.2. The GKS function draw_strings() shown in Function 12.3 is used to draw the results of the generalization, and produced the output in Figure 12.1.

Function 12.1

```
/* generalize a line by N pt retention kcc 6-93 funct 12.1 *
#include "cart_obj.h"
#include "extern.h"
void generalize(number_of_strings)
int number_of_strings;
{
int n, k = 0, str, pt;
STRING gen_line;
printf("Enter n, where every n-th point will be selected:")
scanf("%d%*1c", &n);
for (str = 0; str < number_of_strings; str++) {
  for (pt = 0; pt < line[str].number_of_points; pt++) {
    if ((pt % n) == 0) {
       gen_line.point[k].x = Line[str].Point[pt].x;
       gen_line.point[k].y = Line[str].Point[pt].y;
       k++;
    }}
/* Add the last point */
  gen_line.point[k].x =
  Line[str].Point[Line[str].number_of_points - 1].x;
  gen_line.Point[k].y =
  line[str].point[Line[str].number_of_points - 1].y;
  k++;gen_line.number_of_points = k; k = 0;
  for (pt = 1; pt < gen_line.number_of_points; pt++) {
    Line[str].Point[pt].x = gen_line.Point[pt].x;
    Line[str].Point[pt].y = gen_line.Point[pt].y; }
  Line[str].number_of_points = gen_line.number_of_points;
  } return;
}
```

Function 12.2

```
/*
/* Main program for generalize & draw strings */
* kcc 6-93 function 12.02 */
#include "cart_obj.h"
extern STRING Line[MAXSTRINGS];
main()
{
  int read_strings(), number_of_strings;
  void generalize(), draw_strings();
  number_of_strings = read_strings();
  (void) generalize(number_of_strings);
  (void) draw_strings(number_of_strings);
}
```

Definitive treatments of the line generalization problem are available elsewhere (Buttenfield and McMaster, 1993; McMaster and Shea, 1992). The simplest techniques for reducing the number of points necessary to represent a line are nth-point retention and equidistant resampling. In the first case, the nodes at the end of a vector line representation in entity-by-entity mode are retained as pivots, while for the remainder of the line only every nth point is kept. Similarly, for equidistant resampling, the line is followed by a distance tracking algorithm, and a point is retained at multiples of some key distance along the line.

A number of techniques treat a line as adequately represented by the original data points selected for its representation as a cartographic object, and then fit a smooth curve through the points to make symbolization more attractive or easier to interpret. Splines, polynomials, and Bezier curves are mathematical functions that have been used to perform smoothing. Buttenfield (1985) provided a list of references to the algorithms behind these functions. Many automated contouring packages use these methods to smooth the lines fed though a grid during contouring.

Among the various line generalization methods, one of the most long-standing is the Douglas-Peucker method (Douglas and Peucker, 1973). This method uses the worst-case generalization of a line as the starting approximation, that is, the line segment formed by simply connecting the endpoints. Each point along the line has an orthogonal distance from the line, that can be computed by simple trigonometry. The Douglas-Peucker algorithm selects the point with the largest distance, breaks the line at this point, and then recursively applies this criterion to the resulting segments.

Recursion continues until either only a minimum number of points remain in the string segment, or a tolerance level is reached, perhaps a proportion of the initial orthogonal distance (Figure 12.2).

A number of researchers have suggested and used different criteria for evaluating line generalization methods both subjectively and objectively. McMaster (1986) used a quantitative measure based on the areas of the triangles formed between triplets of points in a line and in its generalization. Muller (1986) has suggested as a criterion the ability of an

Function 12.3

```
/* draw a set of strings using GKS kcc 6-93 funct 12.03 */
#include <gks/ansicgks.h>
#include "cart_obj.h"
void draw_strings(number_of_strings)
  int number_of_strings;
{
Gchar *conn = NULL; Gchar *wstype = "postscript";
Gws ws = 1; Gwlimit win; Gnlimit viewport;
int transformation = 1, i, j;
float aspect_ratio;
/* Open up GKS */ if (gopengks(stdout, GMEMORY)) exit();
/* Open the workstation */ gopenws(ws, conn, wstype);
/* Activate the workstation */ gactivatews(ws);
/* Find the bounding reactangle of the points */
win.xmin=line[0].point[0].x; win.xmax=line[0].point[0].x;
win.ymin=line[0].point[0].y; win.ymax=line[0].point[0].y;
for (i = 0; i < number_of_strings; i++) {
  for (j = 0; j < line[i].number_of_points; j++) {
if(Line[i].Point[j].x<win.xmin)win.xmin=Line[i].Point[j].x;
if(Line[i].Point[j].y<win.ymin)win.ymin=Line[i].Point[j].y;
if(Line[i].Point[j].x>win.xmax)win.xmax=Line[i].Point[j].x;
if(Line[i].Point[j].y win.ymax)win.ymax=Line[i].Point[j].y;
} }
  gsetwindow(transformation, &win);
  aspect_ratio = (win.ymax-win.ymin)/(win.xmax-win.xmin);
  viewport.xmin = 0.0;viewport.ymin = 0.0;
  if (aspect_ratio < 1.0) {/* Landscape */
    viewport.ymax = aspect_ratio;viewport.xmax = 1.0;
  } else {/* Portrait */
    viewport.xmax = aspect_ratio;viewport.ymax = 1.0;};
  gsetviewport(transformation, &viewport);
  gselnormtran(transformation);
  for (i = 0; i < number_of_strings; i++)
    gpolyline(Line[i].number_of_points, Line[i].point)
  gmessage(ws, "quits after 35 seconds");sleep(35);
  /* Deactivate the workstation */gdeactivatews(ws);
  /* Close the workstation */ gclosews(ws);
  /* Close down GKS */ gclosegks();
}
```

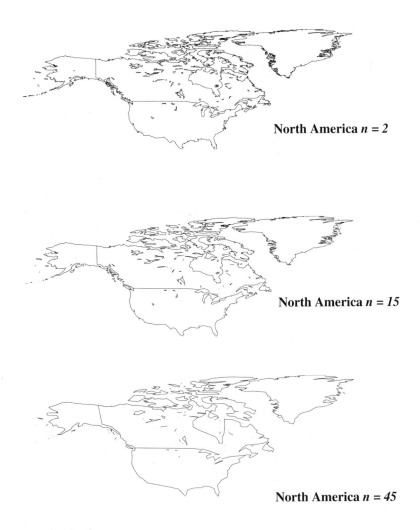

Figure 12.1 Line generalization by *n*-point retention.

algorithm to preserve the fractal dimension of the line. The fractal dimension, a value reflecting the degree of scale invariance of a line, seems related to line complexity.

Although line generalization by resampling reduces the number of points in the cartographic object representing the line or in its symbolization, several authors have devised methods to actually increase the number of points. Dutton (1981) proposed an algorithm that computes midpoints for segments and then moves the midpoints using the values of four controlling parameters. Dutton pointed out that fractalization allows lines to have features exaggerated and allows the introduction of smaller scale features with enlargement. The introduced features, being self-similar, have similar properties to the existing line.

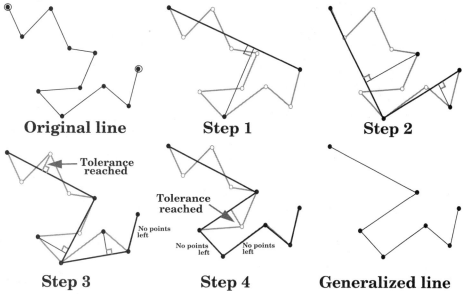

Figure 12.2 Douglas-Peucker line generalization.

12.2.3 Area-to-Area Transformations

Resampling areas to yield areas is a resampling transformation of considerable value to analytical and computer cartography and to GIS. A general statement of the goal is to merge multiple data sets into a set of regions such that the merged data allow comparison between maps. Usually, at least for sets of geographic areas, this means computing a set of greatest common geographic units, or areas that need no longer be partitioned to represent spatial data as cartographic objects.

As a practical expression, consider nonnested regions such as census tracts, school districts or police districts. A crime study may wish to collate population characteristics by police district, only to find that the boundaries do not coincide with census tracts. Similarly, when data have to be compared between different time periods, invariably change in the geographic extent of regions has taken place. How analytical and computer cartography deals with this problem is largely a function of the data structure used to store the various map layers and the data structure in which the map comparison is to be conducted. Clearly, for two maps in two different structures there are two strategies.

First, we can transform both maps into a common data structure that allows comparison directly. Often this is done by converting to a set of topological or entity by entity most common geographic units or to a grid. In the case of the grid, we actually perform

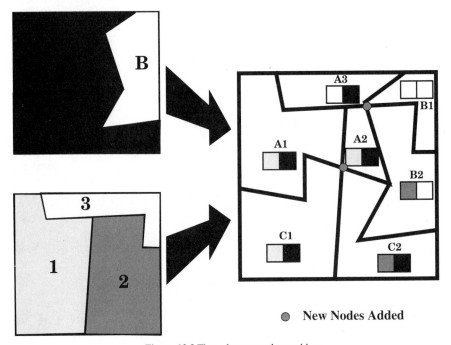

New Nodes Added

Figure 12.3 The polygon overlay problem.

an area-to-point transformation, so that the points coincide for two or more sets of regions. The second strategy is to convert one of the maps into the structure of the other. Thus a polygon entity-by-entity map can be gridded to compare with another grid. As the number of layers increases, the first of these strategies becomes more suitable, especially when inverse transformations are required.

Overlaying two maps to generate a set of most common geographic units in vector mode is not a trivial task (Goodchild, 1978). Central to the problem is the processing of line data to determine intersection points between overlapping polygons so that they can be added to both polygons in the correct place. The points then become nodes in the network of lines that constitute the most common geographic units. A summary of the contributions of computational geometry to these problems is the book by Preparata and Shamos (1985). By the late 1970s, polygon overlay was available within a number of GISs and automated mapping systems (Franklin and Wu, 1987), especially within theWhirlpool module of Odyssey (White, 1978).

The procedure for polygon overlay, illustrated in Figure 12.3, consists of three separate subproblems. First, the intersection points between lines must be found. Usually, this is done using a line segment intersection routine such as that shown in Chapter 10, recursively for all pairs of line segments in the two chains. A way to save considerable computation time is first to check the bounding rectangles of the two chains for overlap. If there is no overlap between bounding rectangles, then there is no need to test each line segment for intersection. When intersections are found, the chains must be split, and the intersection point must be labeled as a node and included as the endpoint of the new sub-divided chains. Many mapping systems compute and save the map coordinates of the bounding rectangles of lines and areas automatically on data entry for this purpose.

In the next stage, the partitioned chains are reassembled into the new set of polygons that make up the most common geographic units. Finally, each polygon must be rela-beled, either with a new set of sequential or other labels or with labels that record the par-titioning history. Particularly difficult to process are polygons that cross and recross boundaries, and islands. A FORTRAN implementation of polygon overlay was published by Baxter (1976). Numerous improvements and refinements have been reported over the last few years, and the polygon overlay problem remains the topic of considerable work in analytical cartography.

12.2.4 Volume-to-Volume Transformations

Volumetric representation is rare in cartography because there are few truly three-dimen-sional means by which to symbolize cartographic objects. The usual volume-to-volume resampling transformations are changes in the grid spacing associated with grid data structures or changes in a TIN surface representation. Within a TIN, rarely is the set of points originally used to depict the surface changed, because the points are locations of real data observations and the data structure is fairly compact in the first place. When TINs are compared with data in other formats, such as grids, either the TIN is interpolated to a grid, or the data are points and are allocated to TIN triangles using a point-in-polygon test.

Grids are, however, frequently resampled to change size, to retrieve a subset, or to change the grid spacing. Invariably, unless the change is simple, such as taking every oth-er row and column, the new grid is generated by computing the location of the new grid intersections in the coordinate space of the map and then by interpolation from neighbor-ing grid cells. Usually, the four neighbors and a simple unweighted average are sufficient for the resampling.

12.3 VECTOR TO RASTER DATA AND BACK AGAIN

We have seen in the previous chapters that data structures for analytical and computer cartography often reflect the demands of the hardware and software they support. Just as display devices can be categorized as vector or raster, so can the cartographic data struc-tures used by software be characterized in this way. In the past, a broad division was made between raster and vector data structures. Each structure has its advantages and disadvan-tages and has also had its proponents. Since the development of algorithms for efficient data structure transformation, however, the vector-versus-raster debate has become largely irrelevant.

The reasons for transformation are many, and in many automated mapping systems the conversion of raster to vector data, or vice versa, is a major consumer of computer time. The raster structure is a grid format and stores a map as individual pixels on lines similar to a television picture and a satellite image. The vector structure stores maps line by line, feature by feature. The vector format is most suitable for storing as objects and symbolizing the cartographic entities familiar to human thought (Peuquet, 1979). Most cartographic objects represent boundaries, rivers, coastlines, railroads—distinct lines or groups of lines. On the other hand, raster devices operate reliably and flexibly and have advanced data handling capability. For data storage, the manipulation of data in raster mode is quite straightforward.

The repertoire of vector algorithms for cartographic transformations was more quickly developed than for raster algorithms (Peuquet, 1979). This difference, however, rapidly diminished during the 1980s. Normally, spatial data stored in vector formats take less storage space than does their raster mode equivalent. The graphic output devices in vector form are relatively accurate and distinct, but the speed of output depends on the length of lines on a map.

The greatest need for transformation is that automatic scanners produce raster data, but vector digitizing requires human control. Therefore, to capture large data sets for cartography systematically , the conversion from a raster data structure to a vector form becomes necessary.

12.3.1 Vector-to-Raster Transformations

The conversion of vector data to raster or grid form is usually termed *rasterization*. Rasterization involves four distinct steps. First, the vector data must be read into the computer for processing. The processing can take place one polygon or line at a time or simultaneously for a whole map. Second, the appropriate scales and map transformations should be applied. Normally, the map projection transformation and any resampling transformation will have taken place before this step is reached; for example, a world map may have been transformed to a Mercator projection, clipped at 84 degrees north and 80 degrees south, and the map resampled using the Douglas-Peucker method.

We can now assume that the alignment of the grid will be to the axes of the coordinates used by the input data. The only remaining decision is how many pixels the map will be converted into, and this is determined either by the desired map resolution or the number of rows and columns. Rectangular pixels are sometimes desired, especially to meet the demands of a particular display device, algorithm, or application. This step establishes the geometry of the rasterization. The third stage depends upon technique. In many cases, the grid is partitioned as an array in the computer's memory, and as the points, lines, and polygons are rasterized, the zeros stored initially are changed to ones, or to an index number for the line or polygon.

For very large arrays a second method is used. In this method, the locations of pixels with value one, with their indices if necessary, are stored, either as a file or internally in [row, column, index] format. These data are then sorted by row and column, and an array is constructed by reading the sorted data, filling rows and columns either with the stored value or with a zero. An option is to fill a polygon with an index value. The final step is to store the array in the required format, such as with run-length encoding or as a bit map.

Several steps can speed the rasterization process. First, if the distance between two points in the original coordinate system is smaller than half the grid spacing, the second point can be eliminated. This reduces unnecessary data and saves processing time. A careful choice of sorting technique can also make a large difference in the processing time used.

Another step is to process the vector data, computing and storing linear equations of each two points. The purpose of the step is to compute in advance the linear equation constants for each line segment. These can be stored as real numbers, although Bresenham (1965) has shown that only integers are necessary. One problem is the special case of vertical lines, that have an infinite gradient and tend to be common on maps. In this case, the programmer can store the fact that the line is vertical.

The rasterization is now complete, yet for symbolization, the line often needs to be thickened. Simple thickening can parse the whole array and change from zero to one (not line to line) any pixel that borders a line pixel. Unfortunately, this tends to fill in the fine details of wiggly lines. Another technique, which in some display devices can be performed in hardware, is to fill in these neighboring pixels with values to be displayed at a lower light intensity than the line pixels themselves, a technique known as anti-aliasing. Anti-aliasing removes some of the effect of stair-stepping along diagonal lines in raster mode, an effect usually called the "jaggies."

Alternatively, an algorithm that has produced thick lines will be thinned so that all raster lines are of a constant width of one pixel. The degree of fatness of the lines depends on the character of the line, the orientation of the grid, the resolution of the grid, and the algorithm used. For example, in the case of a polygon being rasterized, we could assign an edge pixel to any two or more polygons depending on the algorithm in use. If edges of polygons are treated as lines, then a significant number of pixels may be edges, rather than taking on the attributes of a polygon. If the edge pixels are to be assigned to the polygons, we can use a "dominant" criterion (the polygon owning the majority of the pixel area takes the pixel), a "center point" criterion (the polygon covering the pixel center takes the pixel), or any of several other criteria. Line thinning after the fact can be dependent on the processing direction through the array and other factors.

The algorithm `rasterize` contained on the companion disk to this book probably uses the simplest approach. The earlier function `read_strings()` is used to read the North America data set. Function `extremes()` computes the bounding rectangle for the strings, useful for computing the grid characteristics.

Function `get_gridspecs()` prompts the user for two points defining the grid extents and the *x* and *y* grid spacing. This function computes and returns the sampling distance appropriate to the grid. Sampling at any distance greater than this runs the risk of missing cells along string segments. A loop over all strings first computes the straight line equation of the line segment, testing again for vertical segments, and then applies the sampling into the grid apparent in Figure 12.4. Finally, the grid is written as an ASCII array of ones or zeros. Display of these grid files are covered later in the book.

Output from the function is shown as Figure 12.5. The North America outline, unprojected into a map projection, was gridded at three different resolutions. Notice in this figure how the level of detail decreases and the number of fat lines increases as the cell size becomes larger. Also notice the tendency for lines to be rasterized toward the cardinal directions, especially when they are only slightly diagonal, a good example being the U.S–Canada border.

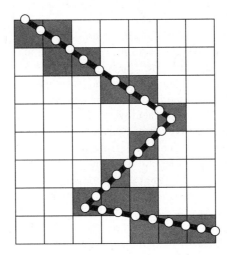

Figure 12.4 Algorithm `rasterize`. String is sampled along segments at half the grid resolution, and the resulting points are converted to integer grid indices. See the `Rasterize` directory of the companion disk.

12.3.2 Raster-to-Vector Transformations

Conversion of vector data to raster is comparatively simple compared with the inverse transformation. The demand for the raster-to-vector transformation is increasing as raster input devices such as scanners and remote sensing devices become more widespread. As a result, the raster-to-vector transformation is being built into automated mapping systems and GISs. This process is very CPU intensive and can yield topological and other errors in the resultant vector data set regardless of the quality of the input data. Analytical cartography has been slow to research issues related to this important transformation, and as a result, the published work is mostly in image processing. Peuquet (1981) noted the lack of efficient algorithms and poor computer coding in this area, a deficiency that has not been fully addressed.

Converting from raster to vector data involves three major steps. These stages, shown graphically in Figure 12.6, can be termed *skeletonization* or *line thinning, line extraction,* and *topological reconstruction.* Line thinning is necessary because vector lines are purely geometric, that is, they have zero width as cartographic objects yet have a width determined by the grid cell size when they appear in raster mode. The grid representation of the lines, therefore, must first be thinned so that the line consists merely of a one-cell-width sequence of connected pixels. This connectivity is usually in one of the eight major directions dictated by the eight-cell connectivity of grid cells.

The first stage, skeletonization, can be performed in one of three ways. Peuquet termed these *peeling, expanding,* and the *medial axis* methods. The medial axis method is the fastest, but it works on one line at a time and not on the whole grid simultaneously,

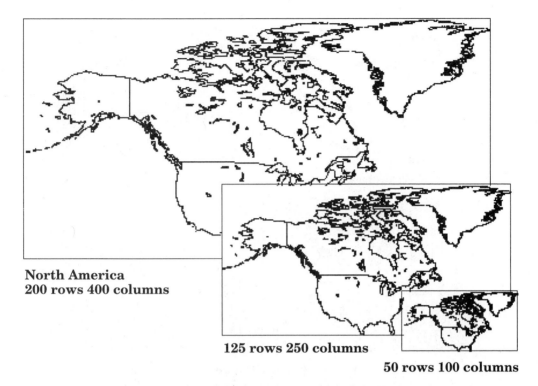

**North America
200 rows 400 columns**

125 rows 250 columns

50 rows 100 columns

Figure 12.5 Algorithm `rasterize` applied to North America.

and inconsistencies occur in very thick lines. Peeling and expanding methods are the inverse of each other. Peeling deals with systematically eroding the edges of lines, wheras expanding deals with expanding the space between lines.

A highly efficient peeling method is Pavlidis's asynchronous thinning algorithm (Pavlidis, 1982). This technique is not completely parallel in operation, however, and in order to maintain connectivity accepts lines with a width of more than one grid cell. Rosenfeld and Kak's thinning algorithm (Rosenfeld and Kak, 1981) is as fast as asynchronous thinning and results in a line of single-grid-cell width. This method works iteratively, at each pass deleting border grid cells that do not disconnect the local (three by three cell) neighborhood. The result of this first step is a single-cell-width line, but sometimes the thinning process introduces artifacts into the geometry of the lines.

Line extraction, the second stage, involves determining where individual lines begin and end in the thinned image, so that the lines can be rewritten as vectors connecting the endpoints in the correct sequence. Points are usually generated at the centers of the line grid cells, and long straight sequences can be eliminated to reduce the number of points in the line. Topological reconstruction is the process of generating the topological connectivity of the lines to rebuild a topological definition of the lines and polygons from the entity-by-entity objects that come from the previous two stages. As such, the third step is identical to the transformations from entity-by-entity to topological data structure discussed in the Section 12.4.

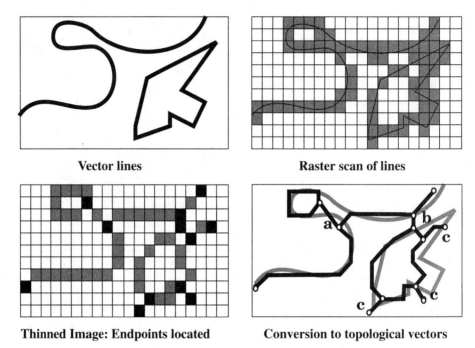

Figure 12.6 Stages and problems of raster-to-vector conversion.
Point a: collapsed loop. Point b: false link. Point c: collapsed spike.

Line extraction can be accomplished in one of two ways. Line following seeks a node that forms the beginning of a line and attempts to follow the line to another node. One of its disadvantages is that when a node is reached and the line terminated, no use is made other lines normally beginning at this point. The scan-line approach uses the same logic as line following, but it processes multiple lines simultaneously.

Once the lines have been extracted, they are usually written as vector lines, and any topological processing takes place in vector mode. The writing of vectors involves finding the absolute or world coordinate location of each grid cell. This can be accomplished by taking the grid cell row or column number, dividing by the number of rows or columns respectively, multiplying by the size of a grid cell in that direction, and adding the world coordinate of the lower left corner of the map. When the grid is not aligned with the coordinate system, the four values stored in the `Grid.corners[]` part of the GRID structure can be used to compute the affine transformation necessary to convert to world coordinates.

Errors associated with the conversion to a finite raster resolution and the use of the various algorithms can result in a significant number of errors, such as collapse lines, spikes, and false links (Figure 12.6). These errors depend on the resolution of the rasters and particularly on the thinning stage of conversion. Where cartographic lines are simple, such as contour lines, this may not be a significant problem. In other cases, such as along a contorted coastline or where contours run close together, the errors can be major and usually require operator intervention. In many systems that do automated line extraction, the conversion process will stop when an ambiguous topological junction is created and wait for the operator to guide the raster-to-vector converter.

When point data rather than vectors are processed, the coordinate transformation is the only one that needs to be applied. A special case is when the data to be processed are coverage polygons without boundaries. Data such as spectral classification and clustering of remote-sensing data , or data from existing raster mode GISs, are often in this format. An approach to finding the boundaries is to scan the grid from top to bottom and from side to side to reveal the boundaries

A simple algorithm to do this is illustrated in Figure 12.7. Alternatively, the grid cell can be filtered using an edge detecting filter, which emphasizes breaks in value. A disadvantage of this method is that one row and one column from each edge of the image are lost for every time the filter is applied. Clearly, when a grid cell is wholly within a polygon, the grid cell is assigned the index for that polygon.

This process is the exact inverse of the way in which polygons are created as grids from vector data. In the vector-to-raster section above, we considered only line rasterization. Points can be gridded simply as individual grid cells, but areas are different. When cells fall into two or more polygons, one way of making the assignment is to compute the area of the cell occupied by each polygon and to assign the cell to the polygon that occupies the most area.

Alternatively, the polygon boundaries can be processed as above and the indices for the polygon interior assigned by filling the bounded area within each polygon. In this case, the boundaries would remain as such and would not be considered part of the polygon. This distinction is made in the Spatial Data Transfer Standards between the ring defining a polygon and the polygon's interior area. For applications where input data are classified remotely sensed imagery, the lack of boundaries is normal, as also is a profusion of small and even single-cell clusters, which would make poor vector equivalents. These can be eliminated by filtering or by assigning small clusters to larger, neighboring clusters if certain criteria are met, such as diagonal connectivity. As remotely sensed data get better resolution, the connecting problem will diminish in significance, although it will remain for small-scale mapping, such as land-use coverage. The elimination of finer detail or "salt and pepper" will become more important with higher resolution.

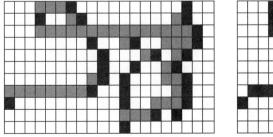 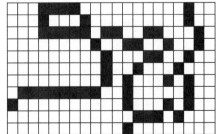

Left to right edges   Top to bottom edges

Figure 12.7 Sweep method for edge location.

12.4 DATA STRUCTURE TRANSFORMATIONS

Data structure transformations are usually necessary because of the demands of a particular mapping system and as such are pertinent to computer cartography. The need for efficient transformationsand the in-depth understanding of the cartographic implications of the transformations, however, are very much a part of analytical cartography. The multitude of data structure transformations performed during the mapping process fall into one of three types. Of these, *scaling transformations* are transformations of cartographic objects by their scale alone. Data structure transformations such as line generalization and grid resampling fall into this category. Resampling transformations, such as the rewriting of cartographic data in the same data structure at a different resolution, are the digital expression of the scale phenomenon. The study of such scale transformations has made important contributions to analytical cartography.

A second type of cartographic data structure transformation, *dimensional transformations,* involves a dimensional change. A dimension change is a change in the type of cartographic object itself rather than a data structure transformation. Such a transformation is a logical data compression. For example, a dimensional data structure change may be the selection of a point to represent an area.

Such a transformation involves considerable loss of data and information, yet gives the clarity or simplicity sometimes required during cartographic generalization. Such a transformation is invertible. We could, for example, generate Theissen polygons from the points to take the place of the original polygons, but the inverse transformation is an imperfect one and results in error. These types of transformations were considered in Chapter 11 as map data transformations,because they involve actual manipulation of the spatial properties of the cartographic objects.

A third type of data structure transformation, *structural transformations,* move cartographic objects between structures as part of the mapping process without much change of scale. For example, a gridded digital elevation model could be processed for significant points such as peaks, saddles, and pits, from which to extract and generate a TIN. The areal coverage is identical; essentially the same cartographic object, the terrain, is available for symbolization or analysis, the object's Euclidean dimension remains the same, yet the data structures are radically different.

The four major data structures we have covered can be classified into entity by entity, topological, TIN, and grid. The grid includes all raster-based structures such as quad trees, and the entity by entity covers some of the hybrid structures, such as Freeman codes. Given these four types of data structures, a matrix of transformations can be compiled that shows the 16 possible transformations (Figure 12.8). The leading diagonal of this matrix involves transformations without a change of data structure and as such these transformations must be *scaling*.

Of the remaining 12 cells in the matrix, four fall into the grid cell to entity by entity and topological structures and their inverses, which were considered above as raster-to-vector and vector-to-raster transformations. The remaining eight cells in the matrix are four transformations and their inverses. These are entity by entity to topological, entity by entity to TIN, topological to TIN, and TIN to grid.

	Entity by Entity	Topological	TIN	Grid
Entity by Entity	Scaling Dimensional	Structural	Structural	Vector to raster
Topological	Structural	Scaling Dimensional	Structural	Vector to raster
TIN	Structural	Structural	Scaling Dimensional	Structural
Grid	Raster to vector	Raster to vector	Structural	Scaling Dimensional

Figure 12.8 Possible data structure transformations.

The first of these transformations, entity by entity to topological, is the normal way by which topology is added during the geocoding process. When maps are digitized, the topology can either be entered explicitly, as in the DIME files, as separate records, or it can be extracted from the points, lines, and polygons as they are digitized. For strings, topological attributes are forward and reverse linkages and the labels of the neighboring polygons.

The linkages to other strings can be determined by storing separately the beginning and ending nodes of the strings, and then matching them against each other. Because manual digitizing results in small differences in the exact coordinate values at points, usually some tolerance is used, and matching points are given the same average location (snapping) (Figure 12.9).

Many digitizing programs maintain and check this information as each string is entered, and these programs prompt the user if an endpoint match cannot be found with the existing strings. The endpoint and topological information can be stored in a structure such as the CHAIN, introduced in Chapter 7.

Finding the names of the neighboring polygons is more difficult, and the usual method is either to prompt the user for the left and right polygon information when a new chain is accepted or to have the user digitize a set of label points, one per polygon. These label points can be stored separately, with the attribute data for the polygons, for example. When polygons need to be referenced, or when attributes are required in the symbolization transformation, the label points can be tested using a point-in-polygon test for inclusion within a group of chains. The relevant data structures are the CHAIN, the AREA_CHAIN, and the NETWORK_CHAIN.

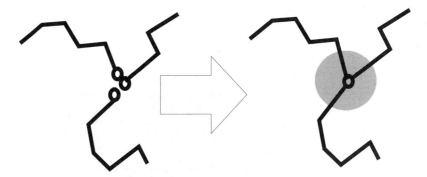

Figure 12.9 Node snapping in entity-to-entity to topological
transformations.

The inverse of this transformation is the extraction of the simpler structures, such as `STRINGs`, `ARCs`, and `RINGs`, from the topological structures. Because the `to_node` and `from_node` information is already available, this conversion is simply a rewriting of the information already contained within the structure. In some cases, chains would have to be written into the `RINGs` in opposite sequence or backwards, that is, from the `end_point` to the `begin_point`, so that the sequential nature of the `RING` is maintained. This is especially important if the motive for the transformation is to symbolize the polygon as a filled area under GKS or if the polygon area is to be computed.

The six remaining transformations all involve the TIN structure. They are topological to TIN, entity-by-entity to TIN, and TIN to grid and their inverses. To date, no real example of a topology-to-TIN transformation is available, although the inverse, in which the TIN is used to generate a stream, ridge line, or connected set of polygons representing visible or invisible areas, seems analytically valuable. This transformation is closer to a dimensional transformation, because the topological data structure is usually two-dimensional while the TIN is three-dimensional. The network, having no partial triangles involved, could be generated using links, with link order being downhill, for streams or ridge lines. For intervisibility problems, however, the splitting of triangles into polygons would be necessary, and simple or complex polygons would have to be formed to store the visible or invisible areas.

The transformation from entity by entity to TIN is also a dimensional transformation, because the only real example of this transformation is the transformation from point entity by entity to TIN, and vice versa. The forward transformation in this case is Delaunay triangulation, while the inverse is again simply a relisting of the entities contained in the TIN structure. Finally, the grid-to-TIN forward and reverse transformation is a true structural transformation, since neither dimensional change nor resampling takes place. This transformation can be accomplished in the forward case by selecting significant points, perhaps by filtering or special case searching, which with their elevations form the basis of the resultant TIN. The inverse transformation is accomplished by linear or other interpolation of elevations at grid cell locations from the TIN's triangles.

This three-type classification of the transformations possible between the four major cartographic data structures can be used to understand the interrelationship among the power, flexibility, efficiency, and storage size of data structures, and it helps determine when data structure transformations are appropriate within a mapping system. Analytical cartography can assist in this understanding by allowing the cartographer quantitatively and theoretically to model and predict the amount and distribution of error to be expected as a result of data structure conversions. The contention is that a cartographic entity is a complex phenomenon, and the means by which this entity is converted to a cartographic object as digital data within a cartographic data structure are both controllable and predictable. With correct understanding of these transformations, cartographers should be able to provide error models and reliability estimates along with maps that are the geographic equivalent of the statistician's normal curve and tests of significance.

12.5 THE ROLE OF ERROR

Much recent research in cartography has been centered on the problem of errors in digital cartographic databases. The focus of the problem is that because digital cartographic databases can be highly precise, many users of the maps produced from the databases also believe them to be accurate. In fact, the data stored as digital maps are often unbelievably faithful reproductions of each and every one of the human errors created in the making of papers maps.

Beard (1989) classified map errors into *source errors, use errors,* and *process errors.* Source errors are errors in the original map sources, in the digital cartographic source data, or in the data capture and geocoding processes. Use errors are caused by a lack of information about a mapping system, unexpected deviation from cartographic convention, the use of maps with a scale which is too small, a lack of timeliness, and a lack of user documentation. Beard proposed that increased attention be devoted to use error, especially because users of GIS are typically not always cartographers.

The Spatial Data Transfer Standard is designed to ensure that the source errors are revealed. The standards uses a "truth in labeling" approach, which also makes the provider of digital cartographic data responsible for providing, either internally or externally, documentation in the form of a quality report. The quality report should contain information on lineage of the data, on positional accuracy of the data, on the accuracy of the attributes associated with the data, and on the completeness of the data. SDTS assists in the task of reducing error by providing a standard set of entity and attribute terms for cartographic features that, if used, reduce ambiguities about digital cartographic data. Also, the standards establish the nomenclature and definitions for digital spatial data transfer, so that data acquired from government or other sources will be in a consistent structure, as documented in the standards.

The final type of error in Beard's classification is process error, the error resulting from the digital conversion, scale change, projection change, or the type of symbolization used. These errors are those attributable to one or more of the cartographic transformations described in this and the preceding two chapters. This type of transformational error clearly relates to the types of transformation discussed above. As we have seen, cartographic transformations can pertain to the map or the data structure, and can involve changes in resolution and changes in object dimension. Three types of process error can therefore occur.

First, errors can occur in the geometry of the map; that is, features can become mislocated in the three-dimensional geometry of the world. These mislocations are sometimes unknown, but can be the result of a controlled distortion, such as changing the map projection or statistical space fitting.

Second, errors can be due to scale change. An ideal situation would be to have a single, very high resolution database from which all others are generated using some objective generalization function (Beard, 1987). This is rarely the case, and maps will continue to be digitized from a large number of sources at different scales. Again, some errors are unknown, such as the removal of islands as they become smaller and smaller with scale, but others, such as the gridding error associated with vector-to-raster conversion, are measurable. In many cases, cartographers have suggested means for defining and measuring errors in these cartographic transformations so that they can be reduced (McMaster, 1986).

Third, errors occur due to the transformation of cartographic objects in digital form between data structures. The case can be made that none of these errors is due to unknowns, because the conversion is under the full control of the cartographer. With the encoding of topology, many consistency checks can become integral parts of GISs and computer mapping systems, ensuring that cartographic errors are more simply detected and corrected (Wagner, 1988). This approach is an integral part of the TIGER files discussed in Chapter 4.

Errors due to changes in data dimension are substantial, but are deliberate in the sense that they allow analysis or display that would otherwise be impossible. It is the straight data structure transformation errors that are most obviously manageable, and part of analytical cartography is clearly the pursuit of data structure transformations that minimize, or at least give complete accounts of, the error they introduce. The standards suggest the use of a quality map, one that maps out the error expected in the cartographic data. Only by fully understanding and modeling cartographic error can cartographers ensure that their products survive the tests of time as today's digital cartographic databases become the historical archives of the future.

12.6 REFERENCES

Baxter, R. S. (1976). *Computers and Statistical Techniques for Planners*. London: Methuen.

Beard, M. K. (1987). "How to Survive on a Single Detailed Database." *Proceedings, AUTOCARTO 8,* Eighth International Symposium on Computer-Assisted Cartography, Baltimore, March 29–April 3, pp. 211–220.

Beard, M. K. (1989). "Use Error: The Neglected Error Component." *Proceedings, AUTO-CARTO 9,* Ninth International Symposium on Computer-Assisted Cartography, Baltimore, April 2–7, pp. 808–817.

Bresenham, J. E. (1965). "Algorithm for Computer Control of a Digital Plotter." *IBM Systems Journal,* vol. 4, no. 1, pp 25–30.
Buttenfield , B. (1985. "Treatment of the Cartographic Line." *Cartographica,* vol. 22, no. 2, pp. 1–26.

Douglas, D. H., and Peucker, T. K. (1973). "Algorithms for the Reduction of the Number of Points Required to Represent a Digitized Line or its Caricature." *Canadian Cartographer,* vol. 10, pp. 110–122.

Dutton, G. H. (1981). "Fractal Enhancement of Cartographic Line Detail." *American Cartographer,* vol. 8, no. 1, pp. 23–40.

Franklin, W. R., and P. Y. F. Wu. (1987). "A PolygonOverlay System in PROLOG." *Proceedings, AUTOCARTO 8,* Eighth International Symposium on Computer-Assisted Cartography, Baltimore, March 29–April 3, pp. 97–106.

Goodchild, M. F. (1978). "Statistical Aspects of the Polygon Overlay Problem." *An Advanced Study Symposium on Topological Data Structures and Geographic Information Systems.* Cambridge, MA: Laboratory for Computer Graphics and Spatial Analysis, Harvard University.

McMaster, R. B. (1986). "A Statistical Analysis of Mathematical Measures for Linear Simplification." *American Cartographer,* vol. 13, no. 2, pp. 103–116.

Muller, J. C. (1986). "Fractal Dimension and Inconsistencies in Cartographic Line Representations." *Cartographic Journal,* vol. 23, pp. 123–130.

Pavlidis, T. (1982). "An Asynchronous Thinning Algorithm." *Computer Graphics and Image Processing,* vol. 20, pp. 133–157.

Peuquet, D. J. (1979). "Raster Processing: An Alternative Approach to Automated Cartographic Data Handling." *American Cartographer,* vol. 6, no. 2, pp. 129–139.

Peuquet, D. J. (1981). "An Examination of Techniques for Reformatting Digital Cartographic Data. Part 1: The Raster-to-Vector Process." *Cartographica,* vol. 18, no. 1, pp. 34–48.

Preparata, F. P. and M. I. Shamos (1985). *Computational Geometry.* New York: Springer-Verlag.

Rosenfeld, A., and A. C. Kak (1981). *Digital Picture Processing.* 2d. ed. New York: Academic Press.

Wagner, D. F. (1988). "A Method of Evaluating Polygon Overlay Algorithms." *Technical Papers,* ACSM-ASPRS Annual Convention, vol. 5, pp. 173–183.

White, D. (1978). "A New Method of Polygon Overlay." *Harvard Papers on Geographic Information Systems,* vol. 6, Cambridge, MA: Harvard University.

13

Terrain Analysis

13.1 THREE-DIMENSIONAL TRANSFORMATIONS

The cartographic transformations discussed so far have avoided instances where three-dimensional data are involved. In analytical and computer cartography, much digital data is three-dimensional, usually relating to the surface of the land or the bottom of bodies of water. These data are three-dimensional because we need eastings and northings *and* the elevation at a location to give the necessary level of information. The part of analytical cartography concerned with the analysis of terrain-type data, including any continuous surface, is called terrain analysis. For this chapter, we have broadened terrain analysis to include the symbolization of terrain using computer cartography. Both analytical and computer cartography must be concerned with the data structures and the special-purpose data structure conversions involved in terrain analysis.

The first of these data structure conversions is by far the most critical and can fully determine the level of accuracy and the error involved in terrain analysis and mapping. This transformation is that of point data to both the TIN and the grid three-dimensional formats. In only a few instances are terrain data collected at locations that are evenly spaced. Usually, elevations are measured at significant points on the landscape, such as the crests of ridges, the tops of hills and mountains, and the bottoms of depressions and lakes.

When data are collected on a grid, such as the elevations taken from stereophoto interpretation using an analytical stereoplotter, there is often a need to treat the data as point data for transformation into a particular map projection or spacing and then to reinterpolate to a grid. The point-to-TIN and point-to-grid conversion is of obvious importance. In a TIN, the irregularly spaced points can be used directly as part of the three-dimensional cartographic object definition, eliminating the interpolation problem. For the grid, however, there is no one simple solution to the interpolation problem, and we often are left making trade-offs to achieve terrain maps that suit a specific cartographic purpose.

13.2 INTERPOLATION TO A GRID

A general statement of the interpolation problem would be the following: Given a set of point elevations with coordinates (x, y, z), generate a new set of points at the nodes of a regular grid so that the interpolated surface is a reasonable representation of the surface sampled by the points. Units for x and y are often in meters, but can also be in degrees, feet, and so forth. Units for the z value are usually meters, rounded to the nearest meter, or feet. Ocean depths are often measured in meters, fathoms, feet, and other units.

In the following worked examples, a test set included on the companion disk is used. This is a data set of elevations digitized from a map of the area surrounding the summit of Mount Everest, a total of 5,112 points. Eastings and northings are in meters on an a local grid centered on the summit area, and elevations are in meters. Many applications use rectangular grids, and sometimes the data values are assumed to lie at the intersection points of the grid rather than at the grid squares. In this case, the grid squares are assumed to "contain" the data value.

To build a grid based on data at these points, we use the neighborhood property of the surface. We assume that the elevations are continuously distributed; that is, there are no sudden cliffs, caves, overhangs, or the equivalent, and that values at any point are most closely related to elevations that are in the immediate proximity. Furthermore, we assume that the influence of any point increases as distance to the point decreases, the so-called inverse-distance method. We can choose to make any interpolated surface fit exactly through the data points given, and we can choose whether to allow the highest and lowest points on the interpolated surface to fall beyond the ranges of the data at the points.

Many methods are available for interpolation. Some of them model the entire distribution of the surface and are considered in Section 13.4. The remainder are local operations; that is they work on small areas one at a time to achieve the interpolation. The first set of methods are the simplest and use inverse distance as their neighborhood model. The methods work by moving from grid cell to grid cell, each time computing an interpolated value for that grid cell. In the following discussion, this grid cell will be termed the *kernel*.

13.2.1 Weighting Methods

Weighting methods work by assigning weights to the elevation values found within a given neighborhood of a kernel. The neighborhood is determined in one of several ways discussed in the following section on search methods. In a computer program, the search method computes the distance of each kernel to each point for every kernel. For a 200 by 200 grid, with 1,000 data points, this means 40 million distance calculations, each of which involves taking a square root and squaring two values and then sorting each of the 1,000 distances to determine which is the closest. Hodgson (1989) called this the "brute force" method and suggested an alternative "learned search" approach in which the points are divided into coarser cells called *sorted cells*. The points are sorted within these areas, and this allows the nearest point information to be retained as the algorithm moves from a kernel to a neighboring kernel, saving considerably on the amount of time required.

Once the points within a neighborhood of a kernel are found, the distances from the kernel to each point are computed and used to weight the elevations at the surrounding points. Mathematically, the formula used is

$$Z_{i,j} = \frac{\displaystyle\sum_{p=1}^{R} Z_p d_p^{-n}}{\displaystyle\sum_{p=1}^{R} d_p^{-n}}$$

where Z_p is the elevation at point p in the point neighborhood R, d is a distance from the kernel to point p, and $Z_{i,j}$ is the elevation at the kernel. The value n is the "friction of distance" that allows very distant points to be penalized with respect to closer points. As n increases, so also does the retention of breaks and extremes in the surface. When n is 1.0 and R is 3, the method is equivalent to linear interpolation over the TIN. Values of n for terrain have varied from 1.0 to 6.0, although many cartographers use a value of 2.0, in which case the technique is called *inverse-squared distance weighting*.

Refinements to the technique involve inserting barriers that the interpolator cannot cross, such as fault lines or coastlines, and using the cosine of the vertical angle between the point and the kernel in the weighting to eliminate the shadowing effect of closer points on the same bearing (Shepard, 1968).

A C language function that reads a file containing data points and creates a grid is included on the companion disk as `inverse_d.c`. The number of nearest neighbors is passed to this function as an argument from the calling program. The neighborhood search, however, is performed unlike those described above. As points are read in, they are assigned their place in the as-yet-empty grid. If two or more points fall into one cell, they are averaged in.

Each empty cell is then subjected to a systematic search. Square neighborhoods around the cell are examined until the required number of points are found. It is possible that more than the required number can be found, because one extra row and column are searched on each side of the kernel in each scan. This algorithm has been found to be highly effective and is far faster than the brute force method. It is implemented in the program `terrapin`, also on the companion disk.

13.2.2 Trend Projection Methods

Trend projection methods are designed to overcome the limitation that occurs because grid maxima and minima can lie only at data points. For a regularly sampled set of data points, it is rare that the grid will exactly coincide with the extreme highs and lows of the actual terrain. In trend projection, sets of points within a region are used. The region is usually determined using one of the search methods discussed in Section 13.2.3. Fits to the surface are made locally, using a mathematical projection method such as a bicubic spline or a polynomial trend surface. This "best fit" surface is then used to estimate the elevation at the kernel. If simple linear trend estimates are taken, often for triplets of points in the neighborhood, the actual value given to the kernel can be the mean of the various estimates, as in Figure 13.1, or the value can be weighted in the general direction of the surface trend.

In rough terrain, the trend projection methods may generate an interpolated surface that is far more textured than the true surface, but when the data are sparse and the need for texture information is high, this method may be useful. Sampson (1978) described in detail the trend projection method integrated into the SURFACE II package, a software system used extensively in the geosciences.

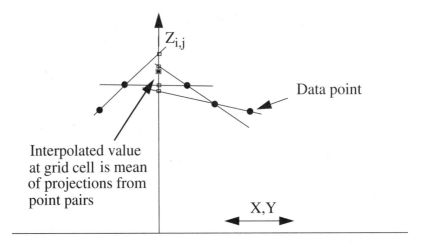

Figure 13.1 Trend projection interpolation.

13.2.3 Search Patterns

Two major variables have a great influence on the two approaches to interpolation discussed above. The first variable is how the neighborhood is chosen. A neighborhood can consist of either a given number of points according to some criterion or of all the points that satisfy certain conditions. The second influencing variable is the relationship between the spacing of the points and the spacing of the grid, and the related problems of the choice of the interpolation area, and the orientation of the grid.

The simplest way of determining which points to use is to include all points in the interpolation. This method, while avoiding the sorting by distance described above, violates the neighborhood property nature of terrain data. Given that we must select points, one approach is to limit the search to points within a certain coarser resolution cell, determined by partitioning up the whole map. This leaves the problem of dealing with the discontinuities that result at the boundaries of cells. The brute force method sorts all the distances and takes for the neighborhood all points that fall less than a given distance or search radius away (Figure 13.2, upper left). This method can result in finding no or few points, especially at the map edges and corners.

An alternative method , which avoids this problem, is to take the nearest R points, regardless of distance or direction (Figure 13.2, upper right). Often, the nearest three, four, or eight points are selected. This method suffers when clusters of points exist, because any interpolation of kernels within the vicinity of the cluster will be overly influenced by the cluster. The problem of directional clustering can be overcome by choosing the nearest points within each of the four quadrants determined by the grid (Figure 13.2, lower left). Finally, each quadrant can be divided to assure equal representation of each octant (Figure 13.2, lower right). Both of these variations suffer from boundary effects; that is,

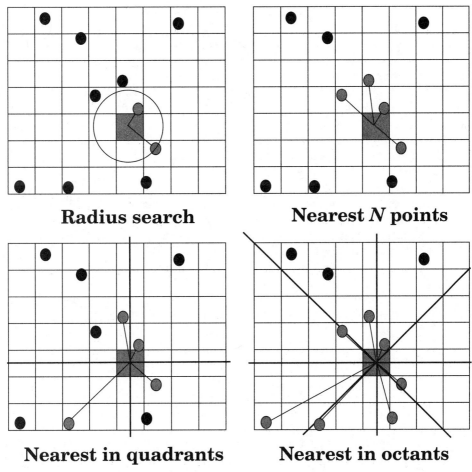

Radius search   ### Nearest N points

Nearest in quadrants   ### Nearest in octants

Figure 13.2 Interpolation point search strategies.

at the edges and corners of the map grid, no points may be available within a quadrant of octant. This problem could be solved by moving to select the points for the neighborhood with another search method when no points are found in the original search area. Clearly, some very complex search strategies can be constructed using combinations of these approaches.

The second of the problems facing the interpolation search strategy is the following: How should the grid relate to the points? The orientation of the grid is usually the same as the coordinate system in use, although this is not a requirement, and for spherical data, where data should be interpolated across the poles, for example, some subtle refinements are necessary. The size of the grid is first determined by the spacing of the grid cells, that is, the grid resolution, and also by the area mapped. The total map area for interpolation is usually rectangular, and the map either is buffered by a data-point void strip around the edges, or it extends over an area for which points lie beyond the map. Clearly, the latter is preferable, because values at the edges of the grid can be interpolated using real data outside the map area, giving the map reliability across the edges. When a void exists at the edge, some interpolator, especially the trend projection methods, can introduce artificial features into the terrain, called *edge effects*.

The grid spacing is also critical. At one extreme, a very large number of evenly distributed points and a coarse grid could allow the grid to be superimposed on the data and all cells to be assigned simply by averaging elevations within cells. At the other extreme, an extremely fine grid could ensure that every data point falls at a grid cell center exactly, leaving the problem of filling in the blanks. In fact, most irregular point distributions contain data-rich and data-poor areas. Statistics like the nearest-neighbor value may be helpful in determining grid spacing; the mean, minimum, and maximum point separation are also useful.

If the source of point data is field measurement, then an effort to get both a uniform spread of points and to collect the data extremes is critical. A map of the distance from each kernel to its nearest data point is a useful aid in determining where to draw the map boundaries and what grid spacing to use. No "best" spacing exists, because the usual limitation is the collection of the data. The sampling theorem states that once a grid spacing is chosen, all features with a spatial size less than twice the spacing are essentially eliminated from the grid, at best becoming random variation or noise. Thus when a grid spacing is chosen, the purpose of the intended symbolization of the terrain should be carefully considered so that features are not eliminated by smoothing.

13.2.4 Kriging

A very large number of methods have been used in the interpolation of point data to a grid. Only one method, however, uses statistical theory to optimize the interpolation. This method, called *kriging*, advanced by Matheron and named for the originator, D. G. Krige, was originally applied to ore bodies in gold mining. Kriging is based on the mathematical theory of the *regionalized variable*. Regionalized variable theory breaks spatial variation down into a drift or structure, a random but spatially correlated part, and random noise.

Thus while hiking up a mountain, the elevation drift is up along a line between the trailhead and the summit, even though we may find local drops to traverse along the trail (random correlated elevations) and boulders to step over while doing so (elevation noise).

Kriging involves a multistep approach to interpolation. First, the drift is estimated using a mathematical function. If none exists, then a good estimate of any map elevation is the mean elevation of the data points. With drift, however, the expected elevation difference with a given separation between the kernel and a point is given by the semivariogram. The semivariogram is computed within regions that are determined using one of the search strategies discussed above.

The semivariogram is then used to statistically fit a model, usually an exponential but a linear model when the semivariogram has no obvious sill, to the distribution. This allows the estimation of semivariance and also the estimation of the weights to be used in computing the local moving average. The weights are chosen so that the best unbiased values are used and so that the estimation variance is minimized. A more detailed discussion, with numerical examples, is contained in Burrough (1986).

Because kriging yields a surface that passes directly though the data points, and because the technique also yields the estimated variance at each interpolated point, the technique is statistically superior to the interpolation methods discussed above. The method is available in several computer packages, even on microcomputers, for example on the SURFER package. The technique of universal kriging works best for data with well-defined local trends. When it is difficult to use the points in a neighborhood to estimate the form of the semivariogram, the model used is not entirely appropriate, and the interpolation may be no better than another method.

While kriging is optimal in statistical terms, it is very computationally intensive and can take up a great deal of computer time, especially for large numbers of data points and large grids. Burrough (1986) has noted few comparative studies of the results of kriging compared with other methods. Figure 13.3 shows the results of gridding the Everest point data set from the companion disk: first using inverse distance squared weighting with a four-point, nearest-neighbor search; and second using universal kriging. In the figure, areas of discrepancy between the two methods are shown in the difference image at the bottom. Kriging estimates are significantly higher than those of the distance weighting in the data-poor zone of the map such as the corners. In other areas, the discrepancies in the distance weighting are over- and underestimates, with the regions of highest slope predominating. Clearly, especially in small grids, boundary effects are major and can hold the balance between differences in the two interpolations.

Research into interpolation continues, and algorithms for gridding are becoming increasingly sophisticated. The major source of data to be interpolated, in addition to field or survey data, is data digitized from contours. Contours are a special case, because they are already graphic interpolations of the point data. In many cases, contour lines contain more information about the terrain than simply the elevations, because when drafting the map, the cartographer often traced features detected on the ground or drew the contours to show streams. Digitizing points from contour maps, therefore, is complex. Simply scanning the contour separation of a map is usually a poor way of generating digital terrain. A more effective approach is to use surface-specific point sampling and then to use either a TIN or an interpolated grid with known interpolation properties to map the terrain.

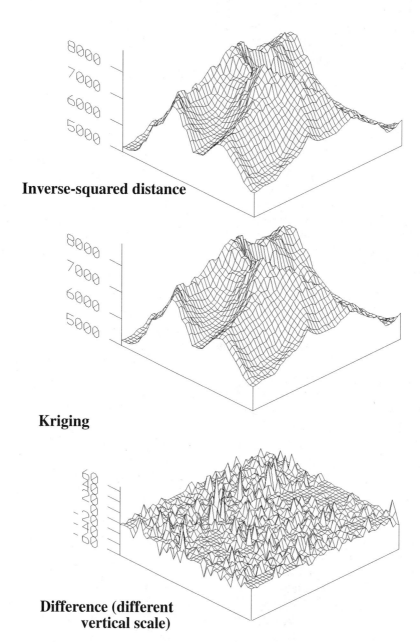

Figure 13.3 Comparison of two grid interpolation methods.

13.3 SURFACE-SPECIFIC POINT SAMPLING

The logic behind surface-specific point sampling is that contour maps, and even field elevation data collected by topographic surveys, contain information about the terrain surface that would be lost by placing a grid over the map and digitizing elevations at grid cells. Significant features that dominate the form of the terrain are streams, ridges, summits, saddle points, and the bottoms of depressions (Douglas, 1986). In addition, actual point elevations are often available on the map as benchmarks, data collection points, or spot heights..

Ignoring the structure or "skeleton" of the terrain, one approach to converting a contour map into either a grid or a TIN is to digitize points along the contours with the attribute of the elevation of the contour. This may work well in only two circumstances: when the point density along the lines is about the same as the map spacing between the contours and in terrain that is very rough. In most cases, more points are digitized along the contours than between them, as shown in Figure 13.4.

When this is the case, interpolation, especially when only a few points are used in the point search, will result in artificial plateaus within the loops of the contours. Since contours usually move back and forth much like a stream, the result is a series of flat terrain steps or plateaus along the contour line, often called the "wedding cake effect" as shown in Figure 13.5. The wedding cake effect is often invisible if the digitized data are only proofed by contouring. It becomes obvious when the resultant grid is hill-shaded.

These steps are artificial and are the result of poor digitizing technique associated with the weaknesses of certain interpolation methods. Avoidance of this problem means taking into account the skeleton of the terrain. Digitizing software which permits an automatic increment by the contour interval with each point digitized is particularly valuable when digitizing streams and ridges. An effective strategy is to start at the low point on a stream or ridge, and to continue up until the high point is reached. Saddle points, summits, and pits (depressions) are then places where the stream and ridge lines converge (Figure 13.6). This terrain structure is sometimes used in environmental modeling to partition the terrain into slopes and watersheds divided by streams and ridge lines, plus features like peaks, depressions and other significant points. Algorithms exist to detect these features in grids, and to vectorize the features for use in GISs.

Points along the skeleton are the targets of terrain data capture. Additional ridge lines can follow significant features up slopes, such as the ridge edges formed by a stream incision. The significant points, such as summits themselves, and other benchmarks, should also be entered. Finally, the intermediate empty spaces can be filled with the occasional point digitized along contours, although because of the wedding cake effect, following a single contour should be avoided.

The result is a set of points which best represent the surface of the terrain. If the TIN were generated from these points, the TIN would be a best-fit model of the surface, and the triangles generated would be good representations of the actual slope facets on the ground. This set of points is also optimal for interpolation, since the interpolator is often better at assigning intermediate elevations from the skeleton than the original manual cartographer was at drafting contours.

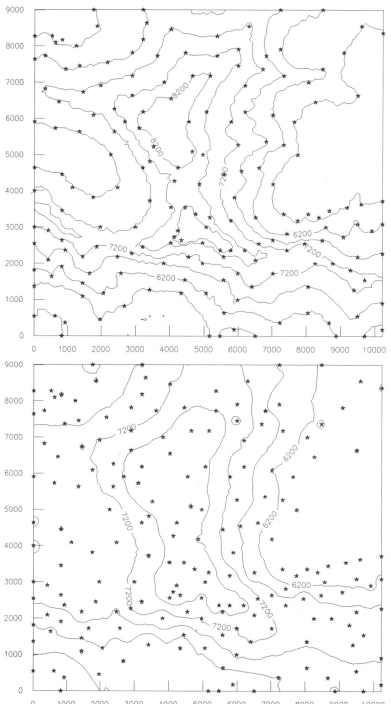

Figure 13.4 Errors in gridding from digitized contours. Top: Gridded from all points. Bottom: Gridded only from the points shown along contours.

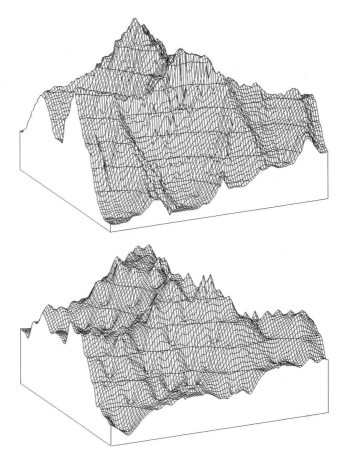

Figure 13.5 The "wedding cake" effect. (Same data as Figure 13.4.)
Top: All points. Bottom: Points on contours only

13.4 SURFACE MODELS

The interpolation methods discussed so far all fit the interpolated surface by local adjustments. Other methods, especially those which for analytical or modeling purposes seek to generalize the surface, use a global or entire surface approach. The *surface models* fall into categories based on the mathematics that describe the surface. The two major surface models, which are global in scope, are polynomial series and Fourier series. Each has analytical powers beyond generalization, and each has been used extensively in analytical cartography.

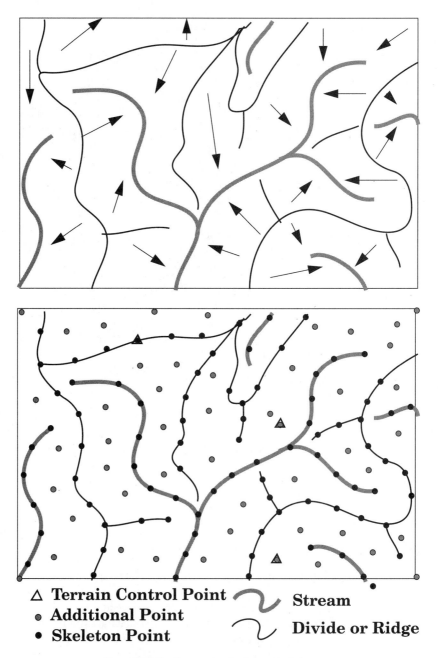

△ **Terrain Control Point**   ∿ **Stream**
● **Additional Point**
● **Skeleton Point**   ∿ **Divide or Ridge**

Figure 13.6 Significant points in the terrain skeleton.

13.4.1 Polynomial Series

Polynomial series are power series in that they are series summations of increasing powers of eastings and northings and their cross products. In the simplest form, a linear surface can be represented mathematically as

$$Z_{x,y} = \beta_0 + \beta_1 x + \beta_2 y$$

As the first constant increases, so the overall average height of the surface increases. As each of the others increases, so the linear plane surface represented by the equation dips or tilts down more and more up to the west or south, respectively. Because a flat tilting plane is rare as terrain, real data seldom fall on such a plane but overshoot and undershoot it. Figure 13.7 shows an artificial surface generated by a computer program. This surface is a cubic surface function, where elevation is a function of x, y, x and y squared, and y cubed. Such a surface can be thought of as having two components, the strongest of which is the overall trend in the data, the drift or trailhead-to-summit path discussed above.

On top of this trend are the local variations, the overshoots and undershoots. These are called residuals: overshoots are positive residuals and undershoots are negative. Statistically, the residuals can be used to fit the linear equation above to a set of points or a surface using least squares. The technique computes the β coefficients for the surface equation which minimize the total of the residuals squared, because some are positive and some negative. The `trend.c` function in the program `terrapin` on the companion disk performs this computation. In the case of Figure 13.7, the β coefficients were:

```
b0 = 177.189765
b1 = 0.871156
b2 = -3.313856
```

The b_1 applies to the x direction, and the b_2 to the y direction. Thus elevation increases with x, but decreases with y, giving the viewing angle of Figure 13.7 as from the north west.

Because trend surfaces are computed by least squares, correlation statistics can be determined. Again, in the case of Figure 13.7, the percent goodness of fit was 99.7 with an R^2 of 0.994. The resultant trend therefore has an associated "goodness of fit" measure, that gives the percentage of the variance in the elevation values accounted for by the linear trend model, which misses only 0.01% of the overall variance.

The fitted trend surface looks almost identical to the original data except that when the pattern of residuals is examined, rounding error (grid cell values were rounded to the nearest whole meter) and the cubic trend become obvious. If the analysis were repeated using a square or cubic trend surface, the fit would have increased to close to one, with the integer rounding error remaining as the sole source of error. This rounding error is not independent of the data, and so it would continue to have spatial pattern and to be related to the relief of the map, that is, the numerical range of the z values. The mapping of residuals has found many applications in geology and geography (Davis, 1973).

When residuals cluster, usually a larger scale spatial process is at work. Increasing the order of the polynomial is one way to capture this variation in the trend surface model. The least squares principle is applied to higher-order polynomials.

For example, a quadratic surface would be given by

$$Z_{x,y} = \beta_0 + \beta_1 x + \beta_2 y + \beta_3 x^2 + \beta_4 y^2 + \beta_5 xy$$

As the surface becomes more complex the number of β terms increases. A cubic surface has 10 terms. At some point, when the number of terms becomes high, the value of the trend surface as a generalization of the original data becomes questionable, because the model has more parameters than we have data.

For cartographic purposes, such as generalizing terrain for mapping or for simplifying slope and aspect computations, a trend surface of appropriate order is adequate. As a theoretical model of terrain, however, the trend surface may introduce unacceptable error during generalization. Davis (1973) provided an excellent discussion of the use and theory of trend surface analysis in geology and published a FORTRAN computer program for its implementation.

13.4.2 Fourier Series

An alternative method for the generalization of terrain is to use a theory developed by the French mathematician Fourier. Fourier showed that any sequence of data could be represented by the sum of a set of trigonometric functions, sines and cosines, over a long enough range of wavelengths. Using space rather than time, because Fourier methods are most commonly used for time series analysis, the wavelength is the distance at which the wave repeats itself. The amplitude of the wave for terrain is the elevation, and the phase angle, the point at which the wave starts, is the elevation at distance zero.

A Fourier generalization of terrain works in two dimensions, x and y, and gives values for elevation. To compute the parameters of the trigonometric functions, Fourier coefficients are computed for all ranges possible within a data set. Because the Fourier analysis of irregular data is complex, the usual data used for Fourier analysis is the grid. Analysis proceeds starting at the longest-wavelength wave that will fit the data, that is, with a wavelength of half the map length in x and half the map length in y. The response or "power" of each pair of x and y wavelengths is computed, a value similar to the percentage of variance explained in least squares. The wavelengths used are the map length L over 2, 3, 4, 5, and so on, until the distance associated with the wave approaches twice the grid spacing, a level at which the Gibbs phenomenon introduces seemingly random errors. This gives a new grid of Fourier coefficients, which are normalized to give the "power" of each pair of wavelengths or "harmonics."

The generalization involves selecting particular pairs of harmonics that are major contributors to the structure of the data. In most phenomena, including linear phenomena such as voice and radio transmission, just a few of the wavelengths carry almost all the structure of the data. These harmonic pairs can be extracted and their Fourier coefficients used to reconstruct a data grid based on their values. This is another example of an inverse transformation.

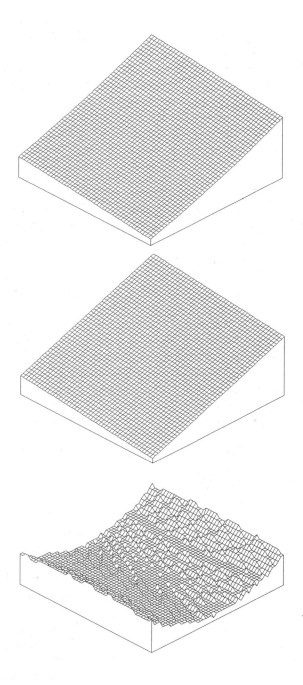

Figure 13.7 Cubic function, linear trend surface and residual (scaled).

   The conversion from the spatial to the frequency or wavelength domain is known as the *Fourier transform*. The inverse Fourier transform can be used to reconstruct either the entire original data grid (less some errors due to boundary problems) or a part of it. A common use of the Fourier transform and its inverse is to eliminate the higher frequencies, that is the variations at small distances. The forward and inverse Fourier transforms give an objective method for this form of generalization.

   Davis (1973) gave a FORTRAN program for performing two-dimensional Fourier analysis. Davis's method used the discrete Fourier transform, a grid cell–by–grid cell method. Far faster is the fast Fourier transform. This technique, however, requires data sets to be of specific sizes. Algorithms, equations, and C language computer programs for both forms of the Fourier transform are contained in Press et al. (1988).

   Application of the discrete Fourier technique to terrain is discussed in Clarke (1988). The byproduct of Fourier analysis of terrain is a precise account of which spatial scales are "active" in a particular piece of terrain. This assists in the choice of a sampling grid as well as in analyzing the processes which have formed the terrain surface.

13.4.3 Surface Filtering

An effect virtually identical to Fourier generalization is possible using local operators. This method is called *spatial filtering*. Filtering is usually performed exclusively on gridded data, and the TIN will therefore be left out of the following discussion. In spatial filtering, the entire grid is processed cell by cell to generate a new or filtered grid. This is done by using a smaller grid, and moving the filter grid step by step, kernel by kernel, over the original data grid. In each case, the filter grid is used to compute a moving average.

   Filter grids are centered on a kernel, and must therefore have an odd number of rows and columns. Filters of 3 by 3, 5 by 5, and 7 by 7 are common. Larger filters lead to problems, because in each application of the filter the original grid gets smaller. Three-by-three filters lose one row from each edge of the grid with each application, five-by-five lose two, and so forth. The filter is placed over the original data grid, and its values are multiplied by the originals to give new values, which are then summed.

   The new filtered value for the grid is then the sum of the values over the filter (Figure 13.8). A requirement for the filter is that the weights or values within the filter grid sum to one. Otherwise, the filter has the effect of damping (< 1.0) or amplifying (> 1.0) the terrain. Filter weights can be tailor-made to yield different effects. Equal weights in each of the cells is simply a moving average or smoothing filter. The weights can be distance weighted by weighting the cells in proportion to their distance from the kernel. A common filter is the Hanning filter, a two-dimensional version of the binomial distribution.

   Special-purpose filters can be designed to enhance features with a specific size, orientation, or characteristic. A horizontal linear filter, for example, can be used to amplify erroneous scan lines on satellite images. A filter with the shape of a ship, for example, can be used to scan ocean images for ships.

   Another commonly used filter is the "lint-picker," a three by three with a weight of minus one in each cell except the center, which has a weight of nine. Note that the weights

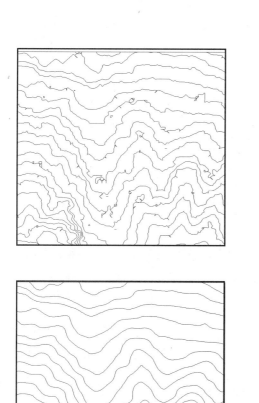

Original Data

125	124	123	121
121	125	125	126
118	120	128	129

1/16	1/8	1/16
1/8	1/4	1/8
1/16	1/8	1/16

Hanning filter

Filter × Original

7.81	15.5	7.69	
15.1	31.2	15.6	
7.38	15.0	8.0	

Summation for kernel

*	*	*	*
*	123		
*			

Figure 13.8
Section of the Everest map
filtered using a Hanning filter.
Top: original data;
middle: filtered three times;
bottom: filtered six times.

in this filter still sum to one. The effect of this filter is to enhance features that are one grid cell in size and are very different from their neighbors. This is a good definition of random noise, so a lint-picker is often used to generate an image or map that is subtracted from the original to remove random noise, the so-called "computer-enhanced image." Filtering an image is fairly easy to program, but can be computationally intensive because the whole grid must be used and a second version of the grid must be saved in RAM during processing.

The function `filter.c` in the `terrapin` directory on the companion disk allows the selection of one of several filters to apply to a grid in the GRID structure. The weights are stored in a vector, and can be adjusted for any size or type of filter with a little modification. The function shown assumes a three by three window and a Hanning filter. With very little modification, the user can place any filter of any size into the `filter.c` function.

13.5 VOLUMETRIC CARTOGRAPHIC TRANSFORMATIONS

Surface models and filtering are transformations of the three-dimensional cartographic data. So far, the motive for transformations discussed have been to transform between data structures (interpolation) and to transform between scales (generalization). Many other transformations are commonly applied to digital elevation data. The purpose of these transformations is purely analytical; that is, the result is a map with enhanced meaning for a specific cartographic problem. Four aspects of analytical transformations will be discussed in this section: the transformation of elevations to slope and aspect, the automatic delineation of terrain-significant points from gridded data, the simulation of terrain data, and the transformations necessary to provide visibility maps.

13.5.1 Slope and Aspect

Evans (1980) noted that slope is defined by a plane tangent to a surface at a specific point and is specified in terms of the maximum rate of change of altitude (slope) and the compass direction associated with the maximum (aspect). Slope, therefore, also has a local neighborhood over which it is computed, usually a three by three region. The maximum slope is familiar to skiers as the fall line, and the aspect is the direction in which the fall line trends. When slope is zero, the terrain is flat and the aspect is undefined. Slope is computed by solving a best fit surface through the points in the neighborhood and by measuring the change in elevation per unit distance in this neighborhood and the direction. These values can be assigned as data to a new grid. Using the TIN, each triangle has a uniform slope within the triangle and also a single aspect. The slope and aspect values in a TIN are discontinuous at the triangle boundaries. Figure 13.9 shows values of slope and aspect computed for an elevation grid covering the summit of Mount Everest. The horizontal resolution is 18 meters per grid cell. The slope map shows a limited set of slopes with extremely steep slopes in white. The aspect image shows with gray tones the eight major compass directions, starting with black for south and ending with white for southwest.

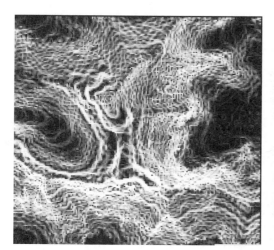

Figure 13.9 Slope and aspect maps for the Everest summit area.

13.5.2 Terrain Partitioning

The primary data structure conversion for three-dimensional data, with the exception of the point-to-grid or point-to-TIN conversion, is the TIN-to-grid conversion, and vice versa. A TIN-to-grid conversion is comparatively simple. Either the points in the TIN can be used in a grid interpolation using one of the methods discussed in this chapter or the elevation values at every point in the grid can be computed using a linear, polynomial, or spline fit to the data in the triangle that includes the point.

The grid-to-TIN conversion, however, is more difficult. There are distinct advantages to performing this conversion. TINs are very compact, are suited to symbolization using polygon rendering hardware and software, and are computationally less demanding than grids. This makes TIN the preferred data structure for solid modeling, for terrain analysis programs, and for surface visualization applications. Essential to the advantages of the conversion, however, is the ability of an algorithm to detect critical points in the landscape. These are the same points discussed as surface-specific in the preceding section. A factor of reduction of 18 times was achieved for conversion of triangles formed by grid points to a TIN based on critical points (Scarlatos, 1989).

The detection of stream and ridge lines in gridded terrain data was discussed by Douglas (1986). Peaks, saddles, and pits are detectable by filtering. The advantages of detecting the significant points is not only for creating the TIN, but also for automatically detecting river channels, for selecting ridges for hill shading and intervisibility, and for dividing the surface into sloping facets for the modeling of hydrology. As yet, however, no simple algorithm to convert a grid to a TIN exists, other than the worst-case solution of dividing every grid cell into two triangles.

13.5.3 Terrain Simulation

Interest has recently been focused on the simulation of artificial terrain with realistic characteristics. Such terrain is used in flight simulators, movies, video games, and models of natural processes. The most common means by which such terrain is produced is using the mathematical methods of fractal geometry. Fractal geometry defines a property called self-similarity, in which as the scale gets larger and larger, geometric form is simply repeated. Two methods for the generation of fractal terrain are commonly used. The first is the midpoint displacement method, in which an original square is drawn, its four corners are displaced either up or down at random, and then the square is divided into four quarters whose corners are displaced by the same amounts, and so on. In each case, the center cell is the computed average of the four corner cells. This method was first used by Fournier, Fussell, and Carpenter (1982) for animations and is fast, simple to program, and gives satisfactory results. For many divisions, however, the technique leaves creases at the edges of the larger cells, for which Jeffery (1987) provided a solution.

A second algorithm, given in Jeffery and attributed to Voss, adds the displacements to all points at each scale instead of just the midpoints. Because this allows steps down of more than a half at a time, the creasing problem disappears. Jeffery (1987) published pseudo-code for his algorithms, and Pascal versions of the programs are available. Information on accessing the programs is contained in an editor's note in the article. Although many fractal surfaces are perfectly adequate for simulations, cartographically incorrect characteristics arise. For example, as many pits as peaks are generated. Often, additional texture algorithms or filtering are used to produce the final "fractal forgery."

13.5.4 Intervisibility

The intervisibility problem can be stated as follows: Given a digital cartographic representation of a terrain surface and a single point on or above the surface, determine the set of regions on the surface that are visible from that point. This set of regions is the opposite of the set of invisible regions. A map of either visible or invisible regions has immense value. Intervisibility maps are used in siting radar and television transmitters, in locating fire towers, in planning ski resorts and housing developments, in highway planning, and in military planning. The solutions to several related problems, such as planning sets of points within view of each other, hiding environmentally obtrusive buildings and land uses in the terrain, and computing the regions where an aircraft is visible to radar, are of direct use in many areas.

The simplest approach to determining intervisibility is to connect a viewing location to each possible target and to follow the line back looking for points that are higher, a method known as *ray tracing*. Any intervening higher point along the ray would screen the target from the viewer (Figure 13.10). Because this involves a very large number of computations, the screening out of invisible areas along rays is advantageous. Intervisibility is easier using the TIN than using a grid. Sutherland et al. (1974) reviewed several algorithms, and DeFloriani et al. (1986) provided a TIN algorithm and Anderson (1982) a grid algorithm. An improved method for the grid was published by Dozier et al. (1981).

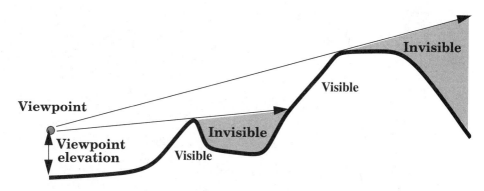

Figure 13.10 Intervisibility using ray tracing.

A primary use of intervisibility algorithms is in selecting viewing locations for gridded and realistic perspectives. The algorithms are also used in calculating the hidden sections of perspective views and in eliminating hidden lines from gridded perspective views. Figure 13.11 shows an application where rays are traced from a viewpoint and are used to assist the cartographer in the selection of a viewing position for generating a realistic perspective view. This step can act as a preprocessor for this computationally intensive task.

13.6 TERRAIN SYMBOLIZATION

The final three-dimensional transformation for cartographic data is that which produces a map. Maps of three dimensional data have problems similar to those with map projections, that is, how to produce on a flat two-dimensional plane a map depicting a volume in three-dimensions. Several cartographic techniques have been devised in recent history for the cartographic symbolization of three-dimensional data. The definitive study of manual methods is Imhof (1982). These methods include contouring and hill shading as well as block diagrams. With computer cartography, many additional methods have arisen, and many methods that required much labor by hand have become available to all.

13.6.1 Automated Contouring

The oldest nonpictorial method of representing three-dimensional data is to use the isoline, first used to show magnetic variation by Sir Edmund Halley. When applied to terrain, the isoline is called a contour and is a line joining points with equal elevation. Reference contours are drawn thicker and numbered at short breaks on the straighter segments. Closed depressions are annotated with hatch marks to distinguish them from hills. Many computer contouring programs automate these features.

Computer contouring can be performed from either a TIN or a grid. Starting at one side of the map, the highest elevation, or the lowest elevation, a first step is to determine whether or not a contour line should appear in a given grid cell or triangle. If the answer

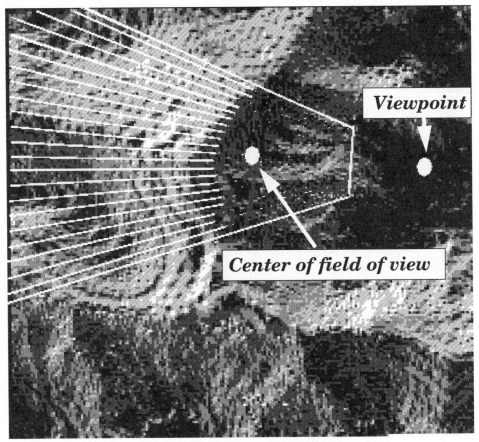

Figure 13.11 Ray tracing for viewpoint selection.

is yes, the next step is to determine points at the edge of the square or triangle where the contour enters and leaves. The next stage uses an interpolator—a quadratic trend surface, an average slope, or Lagrangian interpolation (Crain, 1970)—to generate points along a curve within either the square or the triangle.

An exception in the grid cell case, the case of a saddle point, complicates this process. In this case, a center point is computed as the average of the four corners and is used to move the contour to one side of the cell center. The final contour lines are then smoothed, using weighted averaging or spline functions. A final map consists of a set of points for each contour still structured by grid cell or TIN triangle. In an effort to assure continuity, the contour lines are often resorted so that each continuous loop is drawn without breaking the line, a step that improves plotter output considerably but is unnecessary for many output devices.

As an example, Figure 13.12 is an automated contour map of the Mount Everest summit area, with the exception that the generated contours were smoothed using a different tension factor for the spline smoothing. In some cases, using incorrect values for these coefficients can lead to erroneous contour lines. Some contours become dots or lines at

Figure 13.12 Automated contouring with different spline coefficients for smoothing of contours.

peaks, some cross, and some converge as lines. This last feature is visible on Figure 13.12. An extensive catalog of these line artifacts on automated contour maps is provided by Krajewski and Gibbs (1994). They list 17 types of artifacts commonly resulting from seven variations of six interpolation algorithms and include causes and suggested solutions for the artifacts.

Several inexpensive packages now permit rapid contouring even on microcomputers. It is important to remember, however, just how many parameters go into the production of a computer-generated contour map. The choices of these parameters, data structures, and methods, not to mention the type of output device, are strong determinants of the look of the final contour map. In some cases, the only difference between a manual and a computer contour map is the computer's ability to reproduce the same map given the original data and parameters used.

13.6.2 Analytical Hill Shading

A terrain representation method that has gained considerably in use with computer cartography is hill shading. Automated hill shading works by computing the three-dimensional vector normal to the surface at each point in a grid or for each triangle in a TIN. This normal vector is at right angles to the plane that defines the maximum slope and faces in the direction of the aspect. This vector is projected in three dimensions onto the vector given by a simulated sun angle. The sun, or light source, as multiple sources are possible, is located off the map, with a given zenith and azimuth. The zenith is the vertical angle of the light from the center of the map, and the azimuth is the compass bearing of the sun's direction. If the normal vector faces away from the sun, the surface at that point will be in shadow. If the normal vector points directly to the sun, the point will be fully illuminated. At other angles between the illumination and the normal vector, the point will be partially illuminated.

Analytical hill shading, first derived by Pinhas Yoeli, records the illumination value across the map, usually for every grid cell in a grid, but also over a TIN. Brassel (1974) noted that two values can be used, a reflectance value as discussed above or a density value, which is the logarithm of the inverse reflectance. Figure 13.13 shows the Everest summit elevation data with hill shading using four different solar illumination azimuth directions (NE, SE, NW, SW) and solar zenith angles or elevations of 30 degrees. With control over these angles, it is possible to generate impossible illumination as far as nature is concerned. Hill shading is often used to modify color information on maps derived from remote sensing to present more striking images. Hill shading is also important for generating realistic perspective views.

13.6.3 Gridded Perspectives

The first computer cartographic equivalent of the block diagram used so commonly in geology was the gridded perspective map. These maps are produced exclusively from gridded data, because TINs viewed in perspective appear too complex. Tobler (1970) published a FORTRAN program to generate these views without hidden-line processing.

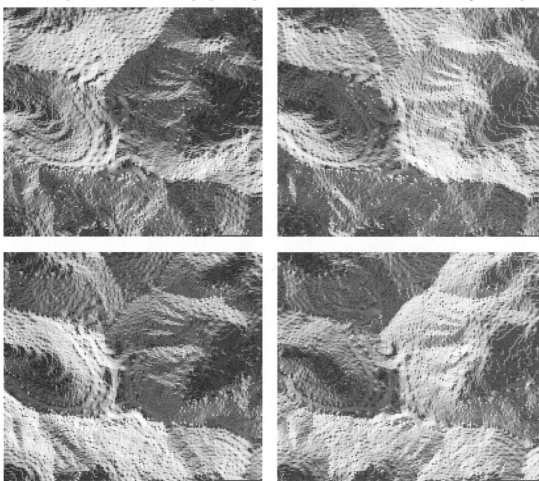

Figure 13.13 Reflectance hill shading with different illumination.

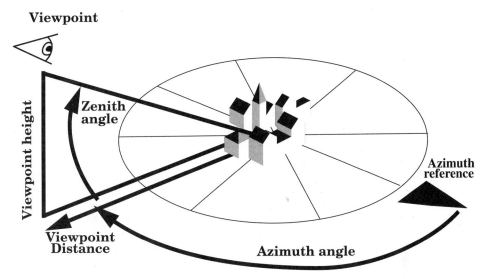

Figure 13.14 Viewing geometry for perspective views.

Many software vendors now sell computer programs to produce gridded perspectives, even on microcomputers with menu control. The critical values in the generation of these images are the viewpoint, the vertical exaggeration, the skirt, the alignment of the lines, and the addition of scales. The viewpoint establishes several things, including the perspective. Views too close have a "wide-angle" perspective, while distant views produce the orthographic perspective favored in block diagrams. Viewpoint can be specified by azimuth, zenith, and distance, by locating a center of the field of view, or by giving a precise triplet of coordinates. The geometry is usually with respect to the center of the volume represented by the image.

The vertical exaggeration is important to the look of the perspective. Textured terrain when exaggerated too much looks chaotic, while too little terrain exaggeration cloaks any actual relief. The skirt is the plain base of the figure. Typically, control over the elevation of the base, as well as whether the surface lines will be drawn over the base, is available. Without a skirt, gridded perspectives seem to "float," but if parts of the underside are visible, added information is gained, especially if the underside is colored differently. Lines on the perspective are usually square to the x and y axes, but variants are to make the lines parallel to the line of sight, aligned to the z axis (raised contours), or sometimes any combination, perhaps in different colors.

The geometry of three dimensional viewing is shown in Figure 13.14. Two variants of fishnet perspective views are stereo plots and anaglyphs, both means by which stereo views can be simulated. In a stereo plot, two images are generated, separated by a 2-degree viewing difference and at a spacing suitable a stereo viewer (Figure 13.15). The separation of 2 degrees between the azimuth angles in the left and right images allow Figure 13.15 to be viewed in stereo because each eye receives a slightly different view, which the brain assembles as a three-dimensional image.

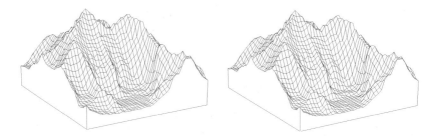

Figure 13.15 Stereoperspective view of Everest summit data.
To be viewed with pocket stereoscope.

    Finally, the anaglyph plot is identical, with the exception that the two different views, with a 2-degree viewpoint separation, are plotted on top of each other. One image is plotted in green or blue and the other in red. Viewing the image through anaglyphic glasses, which have red and blue or red and green lenses, produces the stereo effect. The stereo image, combining two color opposites, should appear as black.

13.6.4 Realistic Perspectives

A refinement of the gridded perspective is the realistic perspective (Figure 13.16). Dubayah and Dozier (1986) presented a summary of work on this method, and discussed algorithms. These images can be generated from both grids and TINs, but employing very different methods. TINs are usually rendered using special-purpose hardware or display software, whereas grids are usually processed in batch mode and the image displayed after completion. The more powerful workstations are capable of almost real-time generation of these images, but even powerful microcomputers may take hours of processing to produce a single image. Hours of supercomputer time have been used to produce sets of these images, which can then be played back in sequence to simulate flight and motion. Most of the same parameters as gridded perspectives apply to realistic perspectives. One major problem is the enlarged effect of grid cells very close to the observer, which can appear blocky. Color for these images is often natural color derived from satellite data.

13.7 TRANSFORMATIONS IN REVIEW

This chapter began with a discussion of cartographic transformations. We have seen that a transformational view of cartography is broad enough to encompass both computer and analytical cartography. Although analytical cartography represents the intellectual challenge to the cartographer, the act of producing the maps often takes up much of the digital cartographer's time. In Part IV, an approach is presented that ensures that this expenditure of time is efficient. To incorporate cartographic transformations as algorithms into mapping systems demands that the cartographer understand both the art of map display and the science of computer programming. Both analytical and computer cartography have much to gain from effective computer programs, just as cartography as a whole can benefit from better and more available maps.

Figure 13.16 Realistic perspective view of Mount Everest.

13.8 REFERENCES

Anderson, D. P. (1982). "Hidden Line Elimination in Projected Grid Surfaces." *ACM Transactions, Graphics,* vol. 1, no. 4, pp. 274–291.

Brassel, K. E. (1974). "A Model for Automatic Hill Shading." *American Cartographer,* vol. 1, no. 1, pp. 15–27.

Burrough, P. A. (1986). *Principles of Geographical Informations Systems for Land Resources Assessment.* Oxford: Clarendon Press.

Clarke, K. C. (1988). "Scale-Based Simulation of Topographic Relief." *American Cartographer,* vol. 15, no. 2, pp. 173–181.

Crain, I. K. (1970). "Computer Interpolation and Contouring of Two-Dimensional Data: A Review." *Geoexploration,* vol. 8, pp. 71–86.

Davis, J. C. (1973). *Statistics and Data Analysis in Geology.* New York: Wiley.

Defloriani, L., B. Falcidieno, and C. Pienovi (1986). "A Visibility-Based Model for Terrain Features." *Proceedings,* Second International Symposium on Spatial Data Handling, IGU Commission on Geographic Data Sensing and Processing and the International Cartographic Association, Seattle, July 5–10, pp. 235–250.

Douglas, D. H. (1986). "Experiments to Locate Ridges and Channels to Create a New Type of Digital Elevation Model." *Cartographica,* vol. 23, no. 4, pp. 29–61.

Dozier, J., J. Bruno, and P. Downey (1981). "A Faster Solution to the Horizon Problem." *Computers and Geosciences,* vol. 7, no. 2, pp 145–151.

Dubayah, R. O., and J. Dozier (1986). "Orthographic TerrainViews Using Data Derived from Digital Elevation Models." *Photogrammetric Engineering and Remote Sensing,* vol. 52, no. 4, pp. 509–518.

Evans, I. S. (1980). "An Integrated System of Terrain Analysis and Slope Mapping." *Zeitschrift fur Geomorphologie,* Supplement-B.d. 36, pp. 274–295.

Fournier, A., D. Fussell, and L. Carpenter (1982). "Computer Rendering of Stochastic Models." *Communications of the ACM,* vol. 25, no. 6, pp. 371–384.

Hodgeson, M. E. (1989). "Searching Methods for Rapid Grid Interpolation." *Professional Geographer,* vol. 41, no. 1, pp. 51–61.

Imhof, E. (1982*). Cartographic Relief Presentation.* New York: DeGruyter.

Jeffery, T. (1987). "Mimicking Mountains." *Byte,* December, pp. 337–344.

Krajewski, S. A., and B. L. Gibbs (1994). "Computer Contouring Generates Artifacts." *Geotimes,* April 1994, pp. 15–19.

Press, W. H., B. P. Flannery, S. A. Teukolsky and W. T. Vetterling (1988*). Numerical Recipes in C: The Art of Scientific Computing.* New York: Cambridge University Press.

Sampson, D. (1978). *Surface II Graphics System.* Lawrence: Kansas Geological Survey.

Scarlatos, L. L. (1989). "A Compact Terrain Model Based on Critical Topographic Features." *Proceedings, AUTOCARTO 9,* Ninth International Symposium on Computer-Assisted Cartography, Baltimore, April 2–7, pp. 146–155.

Shepard, D. (1968). "A Two-Dimensional Interpolation Function for Irregularly Spaced Data." *Proceedings, Twenty-third National Conference, ACM,* pp. 517–524.

Sutherland, I. E., R. F. Sproull, and R. A. Scumacker (1974). "A Characterization of Ten Hidden-Surface Algorithms." *Computing Surveys,* vol. 6, no. 1, pp 1–55.

Tobler, W. R. (1970). *Selected Computer Programs.* Michigan Geographical Publications, Department of Geography, University of Michigan, Ann Arbor.

PART IV
Producing the Map

This book is about cartography, which is about making maps. This part deals with two aspects of the final or symbolization map transformation in which the data and representational techniques are combined with symbols, colors, graphic patterns, and text to produce a map. First, by means of setting the context, Chapter 14 is a brief introduction from the programming practitioner's standpoint to the vast body of knowledge on map design that cartographers have produced in recent years. This chapter is designed to bridge the gap between practice and the literature for the practitioner.

Chapters 15 and 16 deal with how to write computer programs that produce maps. Chapter 15 introduces the issues involved in programming, covers basic software engineering, and uses the GKS programming standard to present a suite of standardized programming techniques for cartographers. Chapter 16 expands upon this to cover language specific issues and discusses how effective user interfaces can significantly enhance the utility of software.

14

Designing the Map

14.1 SYMBOLIZATION IN CONTEXT

A *map* can be defined as a graphic depiction of all or part of a geographic realm in which the real-world features have been replaced with symbols in their correct spatial location at a reduced scale. This definition highlights the importance of the symbolization process in cartography, because the transformation determines not only the esthetic look of the map, but also how effectively it communicates the map's spatial information content to the user or interpreter.

In Part III we saw that the process of mapping can be seen as a series of states at which maps exist in various forms, plus a series of transitions or transformations between these states. The entire sequence of stages begins with the least abstract, with the cartographic and spatial phenomena as they exist in the real world. It continues with the specific steps taken to portray the data that represent these phenomena as objects inside the computer via graphic elements and symbols on a map. The symbolization transformation is unique, first because it produces something tangible (the real map), and second because the tangible map communicates information to its user about the distillation of the part of the world's geography it represents.

A first reference point is the part of the real world that the intended map represents. The term *world* is used loosely, because maps of other planets, moons, and imaginary places are just as cartographic. In the first transformation, a geographic subset of the world is selected, often the area bounded by parallels and meridians or the area defined as a nation, region, or area of interest. Sufficient information about this area is selected from existing maps, or by direct measurement (often from satellite images or air photos), or perhaps by reference to a survey, census, atlas, almanac, or other source of information.

In this initial state, we distill the necessary information needed to compile the map; in other words, we complete the map research and the map at this stage may consist of an existing map, a folder of notes, an air photo, a satellite image, a computer tape, or the correct volume of the census for which we intend to map data.

The first transformation is the process of geocoding, defined and discussed in Chapter 4 (see Figure 14.1). The result is the spatial objects as digital numbers, stored and structured as discussed in Chapters 4, 5, 7, and 8. Attribute data are structured as in Chapter 9. The next transformation is either a map-based or a data structure transformation, as discussed in Chapters 11 and 12. The map data are now ready for symbolization, the subject of this part, and the result is a "real" map. This is not the final stage, however. Significant research in cartography has examined the map-to-map–reader transformation, one that is perceptual and somewhat subjective, but nevertheless important for mapping. This research has led to improvements in design and a better understanding of how people retrieve and then use spatial information from maps.

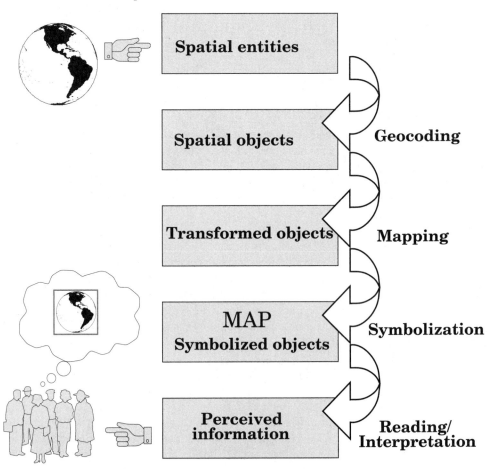

Figure 14.1 States and transformations in a model of the mapping process.

14.2 THE SYMBOLIZATION TRANSFORMATION

Symbolization is itself a set of states and state transformations. The input to the overall process is the map data, transformed into the correct geometry, generalized to the appropriate scale, and clipped to the area of interest. These data define a virtual map, because they could be realized in many different types of symbolizations. The output from the process is a real map, with a selected means of representation or type of map (for example, choropleth or isoline), a given design chosen by the cartographer to facilitate map use, and a set of symbols used to represent the cartographic objects. If the whole set of transformations is completed successfully, the real map will be accurate, effective, esthetically attractive, and useful. The transformation as a whole can be seen in Figure 14.2. Four individual transformations make up the symbolization process: *compilation, representation, design*, and *symbol selection*. Each contributes its own set of modifications to the map data, and represents

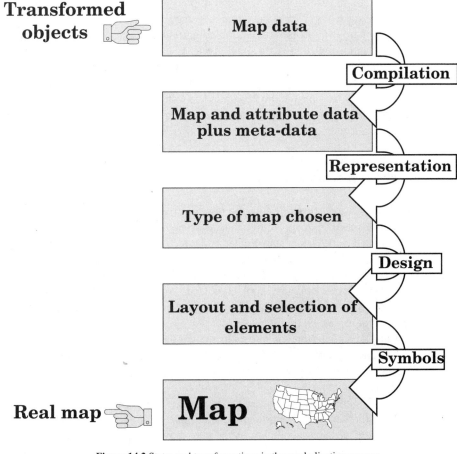

Figure 14.2 States and transformations in the symbolization process.

a successive narrowing in the number of possible virtual maps until just one real map emerges. As the process evolves, the geometry of the map changes from that of world coordinates—in the system of the figure or map area of the map—into the "device" or workstation coordinates of the real map. In the design and symbolization stages, the map space coordinates are used predominantly, and the world coordinates remain only as georeferences into the map figure itself.

14.2.1 Map Compilation

The first of the symbolization transformations is compilation. In this stage, map layers are coregistered at the correct scale and projection, data are selected and converted into the correct form (for example, perhaps county populations are divided by county area to give county population density or objects visible in an air photo are classified), and checks are made to deal with data limitations, missing information, data with different dates, and the reliability of the source information. Selection implies leaving data out, and significant data volume and information loss can occur at this stage.

Compilation involves research including updating and verification, because both the map data and the attribute data should be up to date and correct. The attribute data should be assembled and processed, if necessary, and other metadata, such as the source and date, the type of coordinates and map projection, the data lineage information, and the pertinent accuracy information should be compiled and prepared for use. Primarily, compilation affects how the map data finds a way into the "figure" of the map, one of the cartographic elements considered under design in subsection 14.2.3.

14.2.2 Representation

After compilation, we arrive at the point when we at last have assembled all the information required to make the map. At this stage, the map may be a series of base maps, a set of converted data files, tables, and text data. The representation phase is one in which the cartographer has to choose the type of map that is to be produced. Unwin discussed map types as relating directly to the types of data, in terms of their dimension and scaling. In addition to Unwin's map types of dot, symbol, graduated symbol, network, flow, colored area, ordered area, choropleth, and contour map, cartographers have designed a plethora of different methods for representation.

Two standard cartographic texts that cover design are Robinson et al. (1984) and Dent (1993). These sources yielded a set of representational methods to supplement Unwin's. Figures 14.3 and 14.4 give a sense of how to choose between methods based on the characteristics of the map data.

A basic outline or *reference map* shows the simplest properties of the map data. An example would be a world outline map, with named continents and oceans. A *dot map* uses dots to show the location of features and may show a distribution such as population against a base map. A *picture symbol* map uses a symbol, such a the silhouette of a skier, to locate features such as ski resorts. The *graduated symbol map* is the same, except that the size of

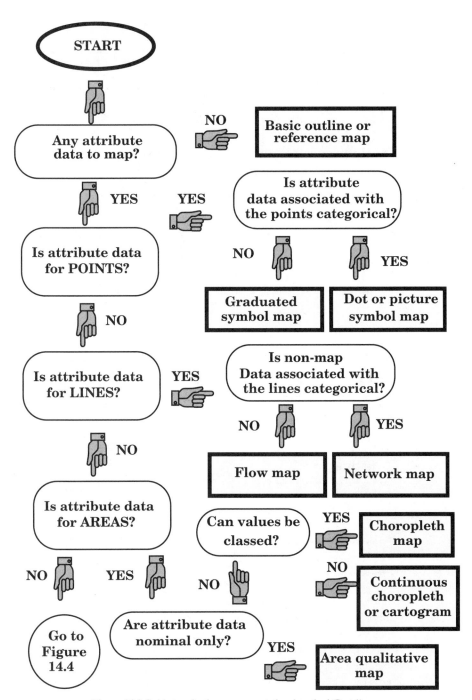

Figure 14.3 Guide to selecting a representational method (Part 1).

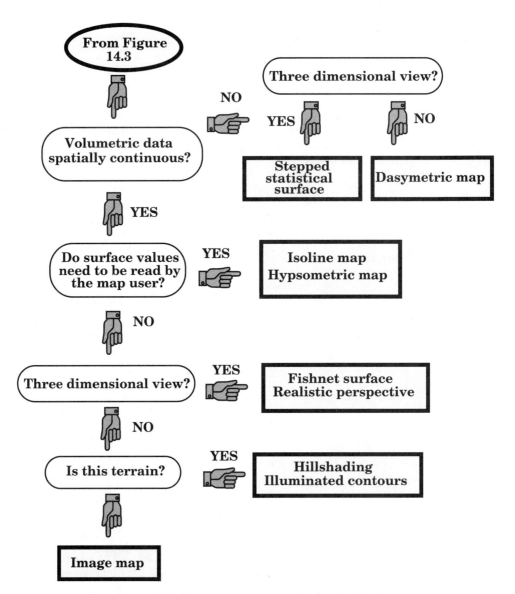

Figure 14.4 Guide to selecting a representational method (Part 2).

the symbol is varied with the value of the feature. Typically, geometric symbols such as circles, squares, triangles, or "spheres" are used.

A *network map* shows a set of connected lines with similar attributes. A subway map, an airline route map, or a map of streams and rivers are examples. The *flow map* is the same, but it uses the width of the line to show value, for example, to show the air traffic volume or the amount of water flow in a stream system.

A *choropleth map* is the familiar shaded map where data are classed and areas such as states or countries are shaded or colored more or less densely according to their value. A variation on this, the unclassed choropleth, uses a continuous variation in tone or color rather than the steps which result from classes. An *area qualitative map* simply gives a color or pattern to an area, for example the colors of rocks on a geological map, or the land-use classes derived from image classification in remote sensing.

Volumetric data can be shown in several ways. Discontinuous data are often shown as a *stepped statistical surface* a block type diagram viewed in perspective. A *dasymetric map* closely resembles an isoline map, without the constraint that values must be sequential. Thus breaks in value are permitted. An example would be a map of rock ore content, which would break sharply at linear fault lines but be smoother elsewhere.

The standard *isoline map* is simply a map with lines joining points of equal value. Surface continuity is assumed, meaning that sharp breaks are usually smoothed. The terrain equivalent is the contour map, with its characteristic datum and contour interval. A variant is the *hypsometric map* in which the space between contour lines is filled with color using a sequence designed to illustrate variation. Often image maps and schoolroom topography maps use this technique.

Three-dimensional views of surfaces rendered in perspective can be either a *gridded fishnet,* as discussed in Chapter 13, or a *realistic perspective,* when an image or shaded map is draped over the surface. The latter technique is often used in animations. Map views of terrain are often represented using *simulated hill shading,* where illumination of shadowing is simulated by the computer and a gray scale or a colored map is used to show the surface. A variant is illuminated contours, in which the shading algorithm is applied only to the contours themselves. Alternatively, apparent hillshading can be produced using inclined contours.

The final map type considered here is the *image map,* in which a value is depicted as variation in tone on a color or monochrome grid. Most raw and false-colored satellite images maps fall into this category, as does the orthophoto map.

This listing is far from exhaustive. A full listing of all representational methods would require an exhaustive survey of the work of thousands of individual cartographers over thousands of years, a virtually impossible task. Cartographers are continuously devising new representational methods, improving the existing methods, and rediscovering methods that are no longer in use. Many of these techniques are attributable to the advent of the digital computer, because their complexity and demands of computation made them impossible to imagine or to implement until the computer came along. Although many of the techniques covered here date back millennia, about half have originated since 1960.

The analytical cartographer should view the representational choice process as a procedure that can be measured, modeled, and improved by automation. Although many of the decisions involved here are subjective, many such problems have been solved using expert systems and neural networks.

One wonders how long it will take before the selection of the mapping method is also fully automated. A computer mapping system can make many of the decisions involved. More difficult, however, will be the automation of map design.

14.2.3 Design

The third transformation in the symbolization process is the conversion of the map data into a map design. Some characteristics of the design are predetermined by the choice of representational method. Primarily, however, the design stage consists of devising a balanced and effective set of cartographic elements with which to construct the map. The interrelationship between the design and the following stage, the choice of symbols, is strong, and it is at this stage where the use of the design loop, a trial-and-error interplay between a map design and a symbol set comes into play. This is because particular symbol sets,such as a set of colors, influence how the map is viewed, read, and interpreted.

A distinct advantage of computer cartography, as previously pointed out, is that this design loop is easily traversed, with a soft-copy or displayed virtual map as the intermediate step. Often, the virtual map is the means by which the design is constructed on-screen in a design software package such as CorelDraw!, Freehand, or Adobe Illustrator. These packages allow the user to preview the map in color, or with its final symbol set, at any time, or even continuously, effectively eliminating the distinction between design and choice of symbols.

Dent (1993) probably gives the most comprehensive overview of map design, oriented primarily toward manual or photomechanically produced maps and thematic cartography. Dent defines the map or cartographic elements as the title and subtitle, the legend, the scale, the credits, the mapped and unmapped areas (figure and ground in Robinson's terminology), the graticule, borders and neat lines, map symbols, and place names and text (Figure 14.5). Dent states that "the task of the designer is to arrange these into a meaningful, aesthetically pleasing design—not an easy task."

Figure 14.5 shows a set of cartographic elements. The *border* is the part of the display medium (paper, computer screen, or other media) that shows beyond the neat line of the map. In special circumstances, additional information can be provided in this space, such as the map copyright, the name of the cartographer, or permissions information. The *neat line* is the visual frame for the map and is usually a bold single or double line around the map that acts as a rectangular frame. From a design standpoint, the neat line provides the basis for the map (that is, cartographic device) coordinate system, in display units such as inches or centimeters on the page.

The two basic parts of the map are the *figure* and *ground*. The figure is the body of the map data itself and is the part of the map referenced in world coordinates. The *graticule* or *grid* (not shown in Figure 14.5) is the reference link between the two coordinate systems on the final map. Also part of both the figure and the ground or legend are the symbols, to be covered in the next subsection. The *legend* translates the symbols into words by collocating text and the symbols in the map coordinate space.

Text information is an integral part of a map, and no map is complete without it. Text figures in the *title* (whose wording sets the theme and the "feeling" for the map), in *place names*, in the *legend*, and in the *credits* and *scale*. The scale is a visual expression of the relationship between the world coordinate space and that of the map space. Because representative fractions change as the map is rescaled into the device space, a graphic scale is preferred. Place names follow a strict set of placement rules, both on the figure and related to features, and within the map space (Imhoff, 1975). These rules have been successfully

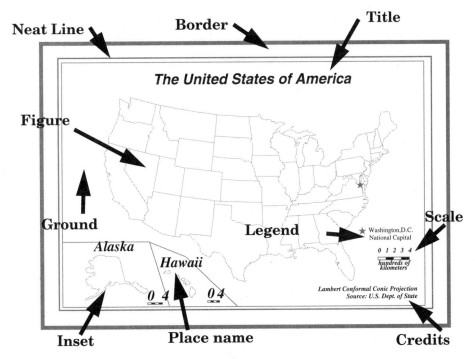

Figure 14.5 Cartographic elements.

automated, but as yet have not been incorporated in computer mapping systems (Ahn, 1984; Doerschler and Freeman, 1992)

Finally, an *inset* is either an enlarged or a reduced map designed to place the map into geographic scale context or to enlarge an area of interest whose level of detail is too specific for the main map scale. An inset should have its own set of cartographic elements, although it is usually highly generalized and many elements may be omitted. Because confusing the figure and ground of the main map with those of the inset is common, the inset should be clearly distinguishable from the figure and ground of the main map. Many Americans, for example, believe that Alaska and Hawaii are islands off of San Diego.

Having named the elements, the next stage is to place them correctly. Actual placement of the elements is usually accomplished in one of two ways: first, by interacting with the map in a visual design loop; and second, by editing a set of macrolike commands that move elements to specific places in the map space. The latter technique is less efficient and involves many traverses through the design loop. Most cartography texts state that the cartographer should aim for harmony and clarity in the composition. This comes from experience and an esthetic sense only and can take years to perfect. For a concise summary of the design literature for the beginning cartographer, the reader is referred to MacEachren (1994).

Obviously, the major facts to bear in mind to balance the elements are (1) that the "weight" of the elements can change when a symbol set (for example, line widths, colors, text fonts and so forth) are chosen, (2) that the elements act in concert with each other in a visual hierarchy (that is some elements naturally stand out from or "above" others, and that using deliberately exaggerated contrast to enhance this hierarchy is usually most effective), and (3) that the combined effect of all the elements is to draw the eye to the center of gravity of the elements. The latter point implies an analytic solution, because the output from many computer mapping systems is a raster file, for which the moment of each pixel around the center of gravity can be computed, along with statistics of trend and dispersal. Cartographic theory implies that the "visual center" of the map should be placed 5% of the map height below the geometric center. Again, computation of the mean and visual center implies an interactive or even fully automated analytic solution to this aspect of map design.

14.2.4 Symbol Selection

The fourth transformation—the symbol selection—changes this map design into a specific set of map symbols, and in this process the cartographer executes the technique to produce the map, with all text, color, linework, and so forth, in final form. The output from this stage is a real map. The map may take many forms and could be a page in a book, a map supplement to a journal, a page in an atlas, a sheet map, an image on a workstation or television screen, a piece of microfilm, a hologram, or a film negative.

The symbolization aspect of design has been studied in detail, and more than a few rules of thumb exist. Some symbolization methods are simply not suitable for certain types of maps and certain map data configurations. For example, a frequent misuse of color is on choropleth maps, especially when the computer gives access to thousands of possible colors (Mersey, 1990). Choropleth maps usually establish value by shading, by pattern, or by color intensity, but rarely by hue alone. Thus a sequence from light to dark red, with a slight hue change looks right, but a sequence of hues from red to blue across the rainbow makes the map look like a decorated Easter egg! Hue changes are appropriate to distinguish between opposites on the same map, such as a surplus/deficit or above/below a statistical average, or two party election results.

Even on general purpose maps, the color balance is essential. Computer displays use pure color, to which the eye is not usually subjected. Less saturated colors, if available, are more suitable for mapping. In addition, convention should be followed. Ground colors are usually white or cyan, not black or bright blue. Contours are frequently brown, water features cyan, roads red, vegetation and forest green and so forth. Failure to follow these conventions is particularly confusing to the map reader. Imagine, for example, a globe with green water and cyan land.

14.2.5 Visual Variables

Most cartographers realize that only a finite set of "visual variables" can be used to generate sets of mapping symbols. This set, first suggested by Bertin (1983), consists of *form, size, color, value, pattern/texture,* and *orientation* (Figure 14.6). Each of the visual variables comes into play during the map symbolization process. The form of cartographic symbols

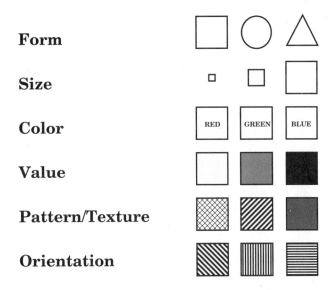

Form

Size

Color

Value

Pattern/Texture

Orientation

Figure 14.6 Bertin's visual variables.

is directly equivalent to the attributes assignable to GKS primitives. Lines can be dotted or dashed, point symbols can be different shaped polymarkers, polygon boundaries can be lines or transparent. The size of symbols can be used as the means of showing data, as in graduated symbol mapping, for example, where again the size is an attribute of the poly-marker. Line size or width is sometimes called weight and is one area where contrast can be enhanced to solve figure/ground problems. Color is a more complex visual variable. Colors are often expressed as red, green, blue triplets (RGB) or sometimes as hue, saturation and intensity (HSI). These values are either integers determined by the hardware device (for example, 8 bit color allows a total of 256 colors from any of $256 \times 256 \times 256$ combinations of individual values of RGB), or, as in GKS, are floating point values of HSI between zero and one. It is possible to translate directly between the RGB and HSI representations of color. Whereas RGB values are simply the degree to which the respective colored phosphors emit light, HSI is closer to the way humans perceive color.

Hue corresponds to the wavelength of light, going from red at the long wave end of the visible light spectrum to blue at the other end. Saturation is the amount of color per unit display area, and intensity is the illumination effect or brightness of the color. Cartographic convention dictates that hue is assigned to categories and that saturation or intensity is assigned to numerical value. When several hues appear in juxtaposition on a map, the colors are perceptually altered by the eye, a phenomenon known as *simultaneous contrast*. Thus maps that use several hues, even as background and line color, should be designed with caution. In addition, the eye's ability to resolve contrast varies significantly with hue, highest in red and green and lowest in yellow and blue.

Value corresponds inversely to saturation and directly to intensity. Because many maps have to appear in black and white, value can be an important visual variable. The eye finds contrasts at low values (near black) much harder to resolve than those at high values. As a result, values are usually assigned to categories or classes. Symbols usually appear with extreme values (for example, white lines on black, black lines on white), but variations can help in setting the visual hierarchy.

Pattern, texture, and orientation are somewhat related, patterns simply consisting of repetitive textures. When patterns are used, they are either conventional (for example, the brick pattern for limestone, the marsh symbol for swamps), or are varied to achieve contrast. In GKS many different patterns are possible using the cell array with duplicated patterns. Pattern should be aligned across features and with the neat line or graticule. A common pattern is the dot sequence to simulate value on printers (for example, in PostScript) and the random or triangular dot pattern (actually a texture) used to reproduce continuous tone in newspapers. Neither pattern nor texture should be used to depict numerical value. Even with classes, a very busy map can result if pattern is varied.

To summarize, the design of a map is a complex process. Good design can be facilitated by planning, by achieving visual balance between map elements, by following conventions, by taking advantage of the design loop, and by the correct use of symbols and map types. Without consideration of design, and certainly without having all the required map elements however impressive it may look on a computer screen, the product is simply not an effective map (MacEachren, 1994).

A well-designed map becomes more than simply an object to be used for communicating map information, it is a reflection of the expertise of the cartographer and the capabilities of the technology of map production, and as such has value beyond its original purpose. The desire to produce computer maps at any cost, or plain ignorance, has often led cartographers to abandon these basic design principles, and as a result we have few beautiful computer-produced maps to show. On the other hand, the means to produce such maps is now easily within reach, if cartographers pay attention to design.

14.3 MAP USE AND ANALYSIS

We now switch from map making to map use. The next reference point is the map reader, who is usually not the cartographer and who may have no cartographic training. The last transformation involves reading the map, and the amount of information lost here can exceed that lost in all the stages up to this. Many cartographers study this transformation, using the term *map perception* to label the stage. This transformation can take place in the field, at night, in a vehicle, or during a location search, and it is usually of great importance in determining the success of the map.

The map-reading transformation uses the real map as source data, either for analysis or interpretation (for example, a farmer measuring the size of fields, or a geologist looking for fault lines) or to simply gain an image in the mind of what it is that the map depicts. If the map is successful, the reader of the map can get locational information as required and without errors, and the interpreter and analyst can do the same. We could say that the information that describes the piece of the world from the beginning of the mapping process has survived the mapping process intact, perhaps it has even been enhanced by the process. In this way, the map can be thought of as delivering a message. If there is only one message, the map is a *thematic map*, that is it shows a single theme. The more information kept as part of the symbolization, the more the map serves the function of storing or preserving generic information. A map showing multiple themes is a *general-purpose map.*

A single pass through the mapping process produces a tangible map and gives it value to the user. As such, the permanent version of the map is a valuable snapshot of the knowledge, skill, beliefs, and objectivity of the cartographer, as well as the technology used in each of the transformations. It is not surprising, therefore, to find map libraries and a healthy trade in old and interesting maps. The study, analysis, and interpretation of historical maps is a very important part of cartography, and is recommended for all those interested in computer mapping. Thematic maps are judged effective if they can communicate to the map reader the fundamental properties and essential characteristics of the particular statistical distribution that they depict. General-purpose maps, however, act as storehouses of locational information and are used for hiking, driving, planning, and even as data sources themselves.

The final transformation in the mapping process encompasses both map reading and map interpretation. Map reading is the inverse of the symbolization process. Locations, text, and symbols are used to rebuild the local geography in the mind of the map reader. The effectiveness of this translation is a result of the accuracy of the map, the legibility of the text, the clarity of the symbols and legend, and the suitability of the chosen mapping technique. Also, however, the effectiveness is the result of the training of the map reader. Map reading can be taught, as it is to navigators, members of the armed forces, foresters, hikers, and planners. Studies show that the more geographic training a person has, the better that person is able to read maps.

In addition, the final mapping transformation includes the analysis of map information. This sometimes involves measurement of location, areas, length, elevations and so forth. Especially important is the calculation of distances and bearing for cross-country movement. Map-related direct measurement is known as *cartometry* (Muehrcke, 1986; Campbell, 1993). Many mechanical devices for map measurement can be replaced in computer cartography if the map data are retained by the mapping system, as in a GIS, although the tools of the map user, the compass, the altimeter, the hand-held Global Positioning System receiver, and the planimeter, for example, will remain in use as long as there are sheet maps.

Last of all, the map user interprets the information from the map. Map interpretation assumes that the user has knowledge in addition to the "where" that is the primary concern of the cartographer. For example, the cartographer may be interested in accurately showing contours and streams in a landscape, while the structural geologist is searching those same contours and streams shown on the map for information about underlying geology, structure, and geomorphology (Miller and Westerback, 1989). The final transformation is not outside of the cartographer's interest, however, since this is the result of a logical and consistent cartographic treatment of the data throughout the mapping process. Cartographers can learn much from using their own and other cartographer's maps in the field.

14.4 THE MAPPING PROCESS AND THE COMPUTER

No part of the mapping process has escaped the impact of the digital computer. We use the computer to assist in map research, map compilation, map design, map production, finding their location in map libraries, and even in writing textbooks on cartography! Although this book focuses on the computer as a means of producing maps, we should not forget that the computer's impact is widespread, from the digital theodolite, to the digital satellite image, to the microprocessor-controlled process camera, to the statistical study of historical maps. This widespread nature of the technology makes the computer unique as a cartographic technology and means that the study of contemporary cartography is impossible without the study of the computer's impact.

We began this book by placing the computer in context as the current technology with which maps are made. This was to demonstrate that the computer is the current logical mapping technology and that its influence upon cartography has been truly revolutionary. Coupled with the data management capabilities of the GIS and linked to the technology of navigation and location-finding through GPS, a highly productive technology-led synthesis of disciplines has taken place, and indeed continues to take place in the mapping sciences.

Balchin (1976) has argued that maps are a component of a graphical form of communication that is a distinct alternative to the more familiar forms of spoken language (articulacy), written language (literacy) and symbolic numerical expression (numeracy and math), which he termed *graphicacy*. Although not everyone uses maps and graphics as a means of reasoning, maps are an extremely powerful tool for the communication of ideas.

The understanding of maps, at least as simple forms, is intuitive to humans, and map interpretation is rapid and efficient. Compare, for example, a text file composed of a regional description (a few kilobytes), a digital sound recording of the same file (maybe a megabyte), and a table of numbers describing the natural and human characteristics of the region (a megabyte at most) with the hundreds of megabytes needed to store imagery and digital map data for a whole region. Yet the user of a GIS can browse through and interpret the spatial information content of the maps in a fraction of the time it would take to read, let alone digest, the text information. Extending this argument, if the goal of a means of communication is to reason and persuade, then no other form of communication can come close to the map. Map logic is a powerful and currently underused tool of communication.

Users of the tools and theory of analytical and computer cartography have the opportunity to play a new role in the information age. Maps have power without doubt. The challenge is to use the power for the benefit of everyone. Effective use of mapping science technology in management and decision making will allow not only the solution of some of the problems that have been traditionally seen as intractable, but also will provide for the more effective use of existing, rather than new, resources.

To use maps effectively for these challenges requires both the practical (computer) and the theoretical (analytical) sides of cartography, for as time advances, cartographic theory becomes cartographic practice. The challenges also require a stream of new cartographers, flexible enough to cross disciplinary boundaries as the technology becomes universal, willing to take up the gauntlet, and intellectually capable of designing and making better maps.

14.5 REFERENCES

Ahn, J. K. (1984). *Automatic Name Placement System.* Publication No. IPL-TR-063, Image Processing Laboratory, Rensselaer Polytechnic Institute, Troy, NY.

Balchin, W. G. V. (1976). "Graphicacy." *American Cartographer,* vol. 3, pp. 33–38.

Bertin, J. (1983). *Semiology of Graphics.* Madison, WI: University of Wisconsin Press.

Campbell, J. (1993). *Map Use and Analysis,* 2d ed. Dubuque, IO: W. C. Brown.

Dent, B. D. (1993). *Cartography: Thematic Map Design,* 3d ed. Dubuque, IO: W. C. Brown.

Doerschler, J. S. and H. Freeman (1992). "System for Dense Map Name Placement." *Communications of the ACM,* vol. 35, no. 1, pp 67–79.

Imhof, E. (1975). "Positioning Names on Maps." *American Cartographer,* vol. 2, pp. 128–144.

MacEachren, A. M. (1994). *SOME Truth with Maps: A Primer on Symbolization and Design,* Washington, D.C: Association of American Geographers Resource Publications in Geography.

Mersey, J. E. (1990). "Colour and Thematic Map Design: The Role of Color Scheme and Map Complexity in Choropleth Map Communication," *Cartographica*, vol. 27, No. 3.

Miller, V. C., and M.E. Westerback (1989). *Interpretation of Topographic Maps.* Columbus, OH: Merrill.

Muehrcke, P. C. (1982). *Map Use: Reading, Analysis and Interpretation,,* 2d ed. Madison, WI: J. P. Publications.

Robinson, A. H., R. D. Sale, J. L. Morrison, and P. C. Muehrcke (1984). *Elements of Cartography*, 5th. ed. New York: J. Wiley.

15

Cartographic Computer Programming

15.1 LANGUAGES

The control necessary for implementing cartographic data structures, for performing cartographic transformations, and for anything other than applied computer cartography, requires the use of a programming language. Many students find their way into analytical cartography from other disciplines and as such are excluded from a more in-depth understanding of analytical cartography because they do not know how to program. Although much of the material presented in this book can be understood, and even taught, without the use of programming, the advanced student may wish to go further. If your background has not been in computing, the best advice is to learn programming from the experts. This usually means in a computer science department. Learning programming along with a graphics course is not recommended.

When we program, we usually do so in a programming language. Just as we use the English language to express our ideas to others, so we use a programming language to express our ideas to the computer. Computer programming languages perform sequences of instructions. This is the case even though our thoughts are often not sequential. A program moves from step A, to step B, to step C, and the move to step C does not start until B is reached. Human thought can easily branch from A to C, or to X, Y, and Z, or all four simultaneously, based on the loosest of connections.

Usually, a program or set of programs written in a programming language performs a specific task. Such a program is referred to as an *applications program*. The other general division of programming is *systems programming*, in which the programs control the operating system itself. All cartographic programming falls into the applications programming category, although there is much that cartographers have to gain from an understanding of systems programming.

All computer programs perform certain generic processes. Most programs perform *input;* that is, they read data from files or a user, they perform *operations* that transform the data in some way (such as between data structures) and then they produce *output.* The output can be a column of numbers, a new file, a graphic, or a map. The most important function of a computer programming language is to control the operation taking place. The programming language gives complete control over an application; it allows us to say exactly what we want to do, how we want to do it, what data should be processed, and where we want to put the output.

Languages have different levels. At the very lowest level of instruction computers use programs called microcode. Microcode is in binary and is so fundamental to the operations of the computer that without it you cannot even load the operating system or use the computer. Microcode instructions tell a computer where to find the operating system, what kind of computer it is, and how to start the *boot* or the initialization process.

At the next level is the machine language program. This language has to be tailored exclusively for a particular brand of computer, a particular type of memory, and so forth. Programs in machine code (language) talk directly to the hardware. Machine code has the disadvantage that not only are data required to be in binary but the instructions are also. The last of the low-level computer languages is assembly language. Assembler is compiled in the same way as higher-level languages. This means that a program can be entered and stored in a file, then processed into machine language via a large translation program called a compiler. The machine language program is then ready to run.

An assembler gives access to *registers* or parts of the memory into which we can write instructions and numbers. Assembly language uses as building blocks an instruction set, consisting of multicharacter mnemonics standing for individual operations. These operations codes perform simple operations, such as putting a number into memory, or taking the current number that is in one register, and performing a binary AND with the number in the current register. Assembler language programs are highly adapted to the architecture of the computer on which they run and as such are capable of running very fast. In many cases, programs are written in higher-level languages, and their most time-consuming modules are rewritten in assembly language for efficiency.

The remaining levels of computer programming languages fall into what are called high-level languages. High-level languages have named operations, expressions that perform control, and defined data types. Examples of widely distributed higher-level languages are FORTRAN (or FORmula TRANslation language) although it certainly was not the only one, BASIC, and COBOL. Others have found their way into general use. A distinguishing characteristic of these languages is that although they contain the mechanics to write modular programs, they do not make the modular structure an integral part of the language.

One step beyond high-level languages like FORTRAN are the structured languages. Structured languages, like Pascal and Ada, encourage the user to write programs that are modular. Modular programs consist of independent units that can be worked on in isolation from each other. A structured language encourages the splitting up of programs into smaller and smaller pieces. Each small program can be worked on, tested, and optimized independently, and then the modules can be assembled into a whole, working program.

Other structured languages are PL/1, ALGOL, APL, C, and RATFOR. In addition, a number of special-purpose languages that are suited to specific tasks, such as text and list processing, and complex database management exist at this level. Among these are LISP, PROLOG, MODULA, SNOBOL, and the object-oriented languages such as C++ and Smalltalk. Beyond the high-level language are application-level programs or macros within systems with their own internal programming language. Graphics languages exist that can perform manipulation of graphic objects and data structures with a command language rather than a programming language. For example, within AUTOCAD, you can program with LISP, or in many systems you can query a database using SQL (Standard Query Language).

The language used throughout this book is a language called C. C is the third version of a programming language developed at Bell Laboratories by Dennis Richie. Originally written on a DEC PDP-11 for the development of the UNIX operating system, the language is general purpose and terse; it supports advanced control and data structures; and it is free of many of the restrictions placed on other languages by their operating environments.

The language was chosen for this book because C language programs are brief, powerful, and support the complex data structures required for cartography and because C is highly portable to different computers and operating systems. The language was originally set forth in a book, *The C Programming Language,* by Brian Kernighan and Dennis Richie (1978). This book set the early C standard (K and R–C), and is the standard followed in this book. Since then, the American National Standards Institute has endorsed a version of C (ANSI-C). The second edition of Kernighan and Richie's book (1988) contains both standards and highlights the differences between them, which to an experienced programmer are minor and are almost entirely limited to type definitions

The C language gains in brevity by leaving out all special-purpose functions, such as mathematical operators, input/output, and text handling, and instead making it easy to build and use program libraries containing working algorithms for these functions. Thus, for example, where FORTRAN defines a SQRT operator to compute a square root, C instead assumes that the user will link the program with a "mathematical library" containing an implementation of an algorithm for computing the square root of a number. C compilers as a result can be relatively small, and many different compilers, debuggers, and programming aids are available for C, even on microcomputers.

In recent years, a preference has been shown for C++, an extension to the C language. C++ adds to the language support for the concept of a *class*, a necessary requirement for the use of object-oriented programming methods. Although object-oriented programming is possible in many other languages, the proliferation of C++ compilers has made C++ the development language of choice, especially because all C++ compilers include the ANSI C language as a subset, with complete forward compatibility. Students should note that effective C++ programming follows from a familiarity with and experience in the C language. Calls to GKS functions are as simple in C++ as in C, because the primary differences relate to how data are modeled (see Chapter 10).

C++ allows the programmer to group together data and operations, such as the actions produced by different functions, together in a class. The graphic object of a polygon encoded as a RING structure, for example, could have a set of functions that relate to it: area computation, point-in-polygon testing, or centroid specification, among others. The complete specification of a set of classes or objects of geographic interest remains to be done. A subset of cartographic objects seems more feasible and is likely to follow the specifications of the DLG-E descriptions advanced by the U.S. Geological Survey.

Finally, C++ allows the specification of complex objects such as windows, user dialogs, and events. This means that C++ is particularly useful when the programmer wishes to use the GUI toolbox directly. Windows and X-windows coupled with C++, for example, allow the programmer access to the building blocks of the user interface with which the operating system itself is constructed, giving significant power of control to the programmer. A consequence, for example, is the ability to inherit the characteristics of the operating system's windows by an application after compilation. More extensive use of C++ is significantly increasing the ability of software to execute in the same way under a number of different GUIs and operating systems.

15.2 GOOD COMPUTER PROGRAMS

Given a group of programmers with different levels of experience and a problem, perhaps even a problem solved by a published algorithm, it is most likely that each programmer will write a different program to solve the problem. Some programs will not work, in which case it is easy to tell whether or not the solutions have value. Among the working programs, however, how can we tell which is the best, or even which are good? To be able to do so requires that we can tell a good program when we see one. Fortunately, good computer programs have some distinguishing characteristics; so much so, in fact, that it is possible to have a good program that does not work and a bad program that does work! In this section, we will try to cover some of the factors that make computer programs "good."

To begin with, a good computer program is *readable*. This means that programmers can read the program, working their way through it without spending an undue amount of time trying to figure out what the original programmer meant. "Clean code" is neatly formatted (usually by a "beautifier program") and is well presented. A good computer program is *structured*. A structured program has embedded within itself subprograms that do smaller and smaller tasks, so the main program consists of very general tasks. This program passes program control to more and more specific tasks, each of which works independently of the others, and returns control when the subtask is complete. This "task nesting" can make programs very manageable and can reduce programming time significantly. Structuring within some languages, such as within FORTRAN, is by function and subroutine. In the C language subroutines do not exist, only functions. So in C a program is defined as a function that is itself a set of functions.

The philosophy behind structuring is that programming complexity is worth minimizing. Writing tiny programs that do very specific tasks is comparatively easy, so if we can take a major task and divide it into many small tasks, each of which can be solved by writing a very simple program, programming complexity is minimized. This approach could be called the "divide and conquer" approach because a difficult problem, divided into less difficult problems, and finally into easy problems, expresses the original problem in a way that can be easily solved. Good programs are structured, and the structure is meaningful in the terms of the problem and works as a solution.

Good programs are *concise,* meaning that they do not contain any more instructions than are necessary to get the task done. A major cause of "verbose" programs is the rewriting performed when program maintenance is done. Rather than reworking the solution, the programmer simply bypasses the old instructions and adds new, duplicating whole sections. Modular programming is by definition concise. When modules are maintained, it is easy to remove a task and rework the solution as a new module or function.

Good programs should be *efficient.* Efficiency is usually measured in terms of the time taken for a task to complete. Many simple measures can be taken to make programs efficient, some of which can make vast improvements in execution time at very little cost. Good programs should be *usable.* Programs that are not usable probably have only one user, the author. Worse than this, once the author is finished using the program it will be forgotten. Modular programs encourage usability because modules can be interchanged. A programmer using structured programming rarely starts a new program from scratch. General-purpose functions can be used and reused in a variety of different programs and contexts.

Good programs are *documented* and *maintained.* Documentation, it is arguable, is more important than the program. Documentation can be internal in the form of headers and comments or external in the form of written text, either interactive (such as help screens) or paged. External documentation should be able to function in a user or tutorial capacity for the new user and as reference for the experienced user. Often this means two manuals, a user manual and a reference manual. Good documentation is understandable; it can be read by a novice user and understood.

Good software is maintained. Good software has somewhere near the beginning a version number, and the version number is high and decimal, such as 4.03. Constant updates mean that a programmer, preferably the author, has gone to the trouble of finding out what is wrong with the software and has changed it to make it better. Usually, for small changes the decimal place is increased. For major revisions of a program, the version number is increased.

Finally, good programs *work.* This means that they produce the correct answer, which is not always the answer we expect or that we want. Working programs must be reliable in two ways. First, they should survive changes in the operating system. Second, programs should be internally consistent. For example, a statistical program should give the same results using the same data more than once. This implies reliability in terms of consistency and in terms of resilience to change.

We have already noted the special demands of cartographic and geographic data. The volume of data that need to be handled influence how we write programs that work with geographic data. Geographic programs also have to run on a variety of different types of computers because no one computer is best for every geographic problem.

Cartographic computer programs very often use graphic output rather than sets of statistics or numbers. Cartographic programs use geographic data structures instead of the data structures that have been optimized for general-purpose data as in database management or numerical computing. Some of the more interesting and challenging data structures that exist are those that are specifically designed to deal with map data. Cartographic programs usually have a very different audience from general-purpose programs.

15.3 SOFTWARE DEVELOPMENT METHODS

15.3.1 Graphics, Standards, and GKS

Graphics are extremely important for analytical and computer cartography. Using graphics with a programming language requires some form of interaction between the program and the graphics. This is accomplished using a *language binding*. There are really two different types of graphics systems, each with different ways of achieving a binding. The first type is extremely device specific and involves a direct or indirect message to the graphics board, the special purpose computer that controls the graphic display. Many graphics programs use the memory addresses on the graphics card directly, and send values to the locations that cause the display to react.

For example, on an IBM PC using EGA graphics in mode 6, memory locations starting at address B800 (hexadecimal) and occupying 640 by 200 binary pixels control the screen display. A binary array written to these locations would automatically appear on the display, in fact, very quickly. We are virtually talking directly to the hardware.

Although very fast, programs that are written in this way usually only work on one type of graphics configuration and on one specific computer. Different computers have different screen sizes, different colors, different memory addresses and address ranges, and different ways of mapping a byte onto bits, and they support different numbers of colors, line thicknesses, and so forth. Not only must each graphics program be rewritten for each display device, it must also be rewritten for each computer. In a rapidly changing hardware environment, to be device-specific means almost instant obsolescence.

As a response, many software manufacturers offer "graphics toolkits," which allow some degree of program portability, but are still very device-specific. Many programming languages, such as microcomputer-based versions of languages like C and Pascal, offer graphics extensions or functions that produce graphics.

At the very simplest, a MOVETO and a DRAW are provided, which allow the user to move to a screen pixel location (the geometry of which is the responsibility of the programmer), with or without drawing a line from the current location. Many toolkits support some sophisticated options, such as graphics text, circles, ellipses, and polygon fills, but the programming level is still fairly close to the machine, and the resulting programs, while a little more portable, are still highly device-specific.

The alternative is to use graphics standards. Two major graphics standards have found their way into analytical and computer cartography. The first of these, CORE, was an outgrowth of a workshop group of SIGGRAPH, the Association for Computing Machinery Special Interest Group on Graphics, in April 1974. The early standard was published in 1977, with completion in 1979. This standard was important in establishing terminology and concepts. The standard was broadly implemented, mostly with FORTRAN bindings, during the 1970s and 1980s. Starting in 1979, an ANSI Technical Committee on Computer Graphics used the CORE system as a model in designing a now well established two-dimensional standard called GKS. Almost all the facilities available in CORE were carried over into GKS and its three-dimensional equivalent, GKS-3D.

CORE showed users the advantages of separating modeling and viewing functions.

This means that a graphical "object" such as a cartographic object, can be defined, given attributes, and so forth, independently of the geometry and properties of the final image, an idea termed *virtual maps* (Moellering, 1983). CORE's weaknesses were the concentration on line-drawing functions, its inability to support a multiuser, workstation environment, and the failure to carry the standard over to include language bindings.

This last weakness meant, for example, that function and subroutine names, the order of arguments in function calls, and even the number of arguments and their meanings were changed by individual software producers. This led to a failure of CORE-based programs to be very portable, one of the principal advantages of using a standard in the first place. The CORE system has not been considered for formal adoption as a standard by any ISO or ANSI committee and will not be in the future because most of its concepts are now embedded in GKS.

The purpose of using a graphics standard is to ensure that a program written using the standard will survive for new systems, will be portable, and will work independently of the specifics of implementation. Using standards, we sacrifice direct access to a device, although GKS provides a high level of specialized access through the generalized device primitive and special escape sequences supported by the language binding. The ability to support such a generic work environment is achieved by separating the graphics into the graphics system, the language binding, and device drivers.

Device drivers are computer programs that convert graphics operations into instructions that communicate directly with the hardware. This requires a device driver for each particular device we are going to use, so that users must add device drivers as new devices are acquired. For the programmer, the trade-off is flexibility for drawing speed. Device independence also means that the same map can appear and look the same on a printer, on an electrostatic plotter, a camera, or a window of a workstation.

As we saw in Chapter 11, with a graphics standard we use three different coordinate systems. We call the coordinate system of the application the *world coordinate system.* Graphics standards use *normalized device coordinates* (NDCs), which are coordinates on a virtual device. A *virtual device* is an idealized planar space on which the map is to be drawn. Under GKS, the lower left corner of the NDC space is at $[0,0]$ and the upper right corner is at $[1,1]$. In the CORE system, which supports three dimensions directly, the center of the screen is $[0,0,0]$, the upper right corner is $[1,1,0]$, and the lower left corner is $[-1,-1,0]$.

At the display end, there are workstation or *device coordinates* which are locations on the display surface as referenced using the mechanics of the display system. Under a graphics standard, the specifics of the device coordinates are unnecessary, although we can inquire their values if need be. In addition, once a window and viewport transformation have been established, the programmer can reference all locations with map coordinates, and specific locations on the map such as locations of legends, titles, insets, and so forth, in normalized device coordinates.

Graphics standards also allow maps to be structured. At the lowest level, the map consists of primitives. Primitives can be collected into *segments*. Segments can be named, numbered, and recalled by reference instead of being rebuilt from data. Different transformations can be applied to whole segments. Similarly, higher-level implementations of the standards support the *metafile*.

The metafile is a generic storage space that behaves the same as a device or workstation. Thus to save the outline of an island, a collection of polylines representing the island could be named as a segment, be written into a metafile, and then later during the same program run, or even later by another computer program, be retrieved and be redrawn with different transformations or attributes. It is sometimes possible to retrieve these metafiles from different programs, perhaps even in different languages.

An additional feature of the standards is the ability to cluster groups of attributes into *bundles*. A bundle can be selected and uniformly applied to either segments or primitives. For example, we could define a set of attributes that should apply to map titles, such as boldface fonts, large letters, a specific spacing of letters, a particular color, and text centering. By referencing this bundle, we can apply it to a specific piece of text that is to be a title, rather than changing all the attributes as we go.

A typical computer graphics program sets the attributes after having computed where it wants to draw primitives, opens a segment, simultaneously defines and draws a segment, and then closes the segment. For example, to draw a solid red polygon first we need the boundaries of the polygon, usually from data. Next we assign characteristics, such as a solid red fill. With this done, we simply open a segment, draw the polygon, and finish. Once drawn and defined, the red polygon can be recalled very easily just by referring to its segment identifier.

In Chapter 16, we will see how to write a graphics program using the GKS standard. The GKS standard has become an important part of graphics programming, especially since the standard's adoption by the American Standards Institute and by the International Standards Organization in 1985. The standard includes a functional description of GKS (ISO 7942:1985), the metafile (ISO 8632-1,2,3,4:1987), and bindings for FORTRAN (ISO 8651-1:1988), Pascal (ISO 8651-2:1988), and Ada (ISO 8651-3:1988). The three dimensional extension of the standard was approved by ISO in 1988 (ISO 8805:1988).

The ISO review of the standard involved over 100 scientists and about a 50 person-year labor effort. The standard has been adopted for several books in computer graphics, including Enderle, Kansy, and Pfaff (1984) and Hearn and Baker (1986). Although other standards, discussed in the following section, are often more suitable for specific graphics and even some cartographic applications, GKS remains the benchmark against which all computer cartographic software in the future will be measured.

15.3.2 Higher-Level Graphics Languages and Standards

Standards in the computer graphics industry have been slow to develop, yet offer a large number of advantages to the programmer. A standard expresses a nominal set of requirements, a minimal level of performance, and a mandatory area of compliance and conformance to accepted specifications (NCGA, 1987).

In the United States, it is the American National Standards Institute (ANSI) that reviews and approves standards and represents the United States in the International Standards Organization. The X3H3 committee of ANSI started examining standards for computer graphics in 1979.

The X3H3 committee initiated GKS, the Computer Graphics Metafile (CGM), and the

Programmer's Hierarchical Interactive Graphics Standard (PHIGS), the Computer Graphics Interface (CGI) as well as GKS-3D.

GKS-3D extended GKS to handle the definition and viewing of three-dimensional "wire-frame" objects. The system supports hidden-line and hidden-surface removal in the workstation transformation, but does not support rendering techniques such as shading and light sources. GKS-3D places a significantly enlarged set of demands upon the host computer, particularly the amount of memory used and the processing power required. As such, microcomputer applications may be limited to only the most powerful machines.

The Computer Graphics Metafile is a file format suitable for saving a "snapshot" of graphics in their final form. Access to images stored in a metafile is either sequential or random access, and the image is defined in a device-independent way. The purpose of the CGM is to allow images generated on different graphics systems to be interchanged and regenerated on different computing and display environments. The three parts of CGM contain the character encoding for assisting in network communications, the binary encoding to support transfer between machines with different word sizes and operating systems, and the clear-text encoding to ensure maximum readability for use and debugging

PHIGS, like GKS, is a functional definition of the links between a programming language and a graphics system. PHIGS supports some operations unavailable in GKS, including the ability to nest graphical objects hierarchically (important in modeling applications and CADD) and graphical object database support. PHIGS is especially useful when complex objects are to be constructed from smaller simpler objects; for example, a fan can be constructed by defining a blade and then repeating it with rotation.

PHIGS supports three dimensions and interactive graphics, such as panning and zooming, at the workstation level. The increase in PHIGS's flexibility and complexity, however, means that many workstations, particularly at the lower end of the power scale, are not capable of supporting all the capabilities of a full PHIGS implementation. PHIGS has come to be used, therefore, in high-end interactive workstation applications, particularly in computer-aided design (CAD), architecture, and computer assisted manufacturing (CAM).

The Computer Graphics Interface (CGI) specifies a set of basic elements for the control of data exchange between the device-independent and device-dependent levels in a graphics system. This involves a standardized virtual device interface (VDI). CGI can operate as software to software, in which case it transfers graphics from metafile to binding, or vice versa.

As a software-to-hardware link, CGI allows stored graphics to be output to any supported device independently of the graphics system which generated them. Thus a map stored in CGI could be reproduced using the available display devices or retrieved for analysis via a computer programming link. The most important function of CGI is to allow device-specific graphics systems to produce output that is in accordance with standards and so to support a larger number of devices.

A large number of additional graphics systems exist at this higher level. For window-based applications, the X-windows system from M.I.T. has gained widespread acceptance. In addition, standards for graphical rendering, and even entirely open graphics-based operating systems, are now in the works. It is important to remember that to be of use a standard should be both supported and used. Programs written to any standard will not have to be rewritten as hardware and software change flavor with the times.

If the early days of computing can be classified as the era of hardware, then the current generation will clearly be the era of software, and above all, programming standards. Links with the standard version of C, now approved at the ANSI level, and with the Spatial Data Transfer Standards, ensure that the duplication of effort and the time wasting of the early days of analytical and computer cartography are over and that cartographers can concentrate upon the more substantive aspects of their discipline, safe in their knowledge of how to produce the map.

15.4 REFERENCES

Enderle, G., K. Kansy, and G. Pfaff (1984). *Computer Graphics Programming: GKS— The Graphics Standard.* Symbolic Computation, New York: Springer-Verlag.

Hearn , D., and M. P. Baker (1986). *Computer Graphics.* Englewood Cliffs: Prentice Hall.

International Standards Organization (1985). *Information Processing Systems—Computer Graphics—Graphical Kernel System (GKS) Functional Description.* ISO Publication No. 7942:1985.

Kernighan, B. W. and D. M. Richie (1978). *The C Programming Language.* Prentice Hall Software Series, Englewood Cliffs: Prentice Hall.

Kernighan, B. W., and D. M. Richie (1988). *The C Programming Language,* 2d ed. Prentice Hall Software Series, Englewood Cliffs, N.J.: Prentice Hall.

Moellering, H. (1983). "Designing Interactive Cartographic Systems Using the Concepts of Real and Virtual Maps. " *Proceedings, AUTOCARTO 6,* Sixth International Symposium on Computer-Assisted Cartography, Ottawa, October 16–21, vol. 2, pp. 53–64.

National Computer Graphics Association (1987). *Standards in the Computer Graphics Industry.* Fairfax, VA: National Computer Graphics Association.

16

Writing Cartographic Software

16.1 THE PROGRAMMING ENVIRONMENT

Writing a computer cartographic program rather than simply using an existing program establishes the degree of control over producing maps necessary for analytical cartography. All too often, however, the decision to write a program is taken too lightly. The result is often confusion, discouragement, and worst of all, a waste of time and resources. With a little thought and planning, however, computer programs that produce maps and map-based analyses can be made to work, and to work well, without the hours of grueling work now known as "hacking."

This development has come about because of some significant improvements in the methods of computer programming. Although programming has been around for many years, only recently has "software engineering," the application of engineering and computer principles to the act of programming, turned the "art" of computer programming into more of a science. Software engineering considers every aspect of a piece of software, from its purpose to its maintenance, to the personnel who will produce it, in addition to the language, coding, and debugging of the program.

The first stage of writing a computer cartographic program is the design. During this stage, we should ask what the software will accomplish, what will be expected of the user, how the software will be documented, maintained, and used, and what its expected lifetime is. Professional software engineers accomplish this step with group meetings and consider a large number of possible alternatives before making selections.

Once the software's purpose, documentation, and user interactions are determined, the next round of decisions should involve a schedule, the choice of a language, a strategy for debugging, a plan for checking and verification, and a plan for the software's use and update. The schedule is important. Programming can lead to large amounts of time being wasted on incorrect solutions, and often the timetable is the only force directing the programmer to give up and try another way. The first draft of a user manual and reference guide, or the writing of on-line help facilities, is a good way to help in planning, organizing, and making design decisions relating to the software. Typically, during the early phase of the design, the software is seen as being able to accomplish too much, so a realistic pruning is often necessary.

In the next phase of design, the software itself is started, but no programs are written. The components are broken down into modules, and decisions are made about how interaction between modules is accomplished. Modules should be chosen so that they are logical divisions of the program, which can be programmed, tested, and maintained in isolation from the program as a whole. At this stage, the places in the program where interaction with the user is necessary must be determined, and the prompts the user is to receive should be planned. Decisions are made about the valid responses to prompts and system defaults that will be taken without the intervention of the user are planned. Finally, the programming language is chosen, and planning begins for the use of the *programming environment.*

The programming environment is the entire set of tools available to the programmer to produce working and correct programs. An ideal environment should include an editor, a debugger, a syntax checker, a beautifier, a compiler, and a linker. The *editor* is the primary tool by which the programmer enters control statements, data, and documentation into the computer. Most people have their favorite editor, and a decision to stay with one's favorite is sound, because many editors have now been converted for use in a large number of programming environments. As a minimum, the editor should be highly interactive, and should allow the partial and complete reading and writing of files, file concatenation (the joining files together), searching for text strings, simple movement through the file, global searching, and substitution (making changes to the whole file). The editor also should be easy to learn and remember. Although allegiance to one editor is one way of remembering commands, the additional features offered by an alternative editor may well be worth a few hours spent reading manuals and running through learning exercises. With the editor, the programmer can enter into a file the C language computer instructions that produce a map, known as the *source code.*

A *beautifier* is a utility program that reads the source code for a computer program and rewrites it with correct indentation, and consistent spacing and with uniform typing specifications. If you have never used a beautifier, it is difficult to appreciate that correct order within a program can be a very effective debugging aid and can even help check for logical consistency. It is not necessary to make programs terse and cramped. Any savings generated by such code can be eliminated by the encoding of a single error that becomes hidden by the sloppy program layout. Many beautifiers are also syntax checkers.

A *syntax checker* is a utility program that checks source code files line by line and detects errors in language syntax, that is, inconsistencies in the rules and requirements of the programming language. A good syntax checker will find errors, point the programmer to the exact place where they were found, and give at least a hint as to which rule has been violated. For example, a common C language syntax error is to forget to terminate a statement with a semicolon. A suggestion by the checker such as "Syntax error in line

23" is vastly inferior to "Missing terminating ';' for line 23". It should be noted that using the compiler to find syntax errors is a waste of time and can lead to a whole new kind of compound logic-syntax error.

The next stage is to *compile* the syntactically correct program. With prior syntax error checking, the compiler is often satisfied fairly easily. This is absolutely no guarantee, however, that the program is correct or even workable. Using C, the final pass of the compiler, called *linking,* has great importance, because C shares program modules between programs. Linking to system libraries such as the math and windows libraries is done at this stage, as also is linking to the libraries that constitute GKS and the GKS-C language binding.

Linker error messages are usually to communicate the fact that a named or required library was not found or that its name was misspelled. After a successful compilation and linking, C usually produces two new versions of the program. The first, from the compiler, is the object module, that is, the compiled machine language version of the program or module. After successful linking, another version, the executable program, is produced. Under MS-DOS, typically a program called prog.c will produce an object module prog.obj and an executable program prog.exe. To run the program, the latter is used.

Often, after all these stages have been completed, the program either produces incorrect results or "crashes." Programs that crash can be started under the control of a *debugger*. Debuggers allow the program to be run, stopped, and the contents of variables listed to determine which variables had what values when or to see which modules were the source of the error. The sequence *editor* to *beautifier* to *syntax checker* to *compiler* to *linker* to *debugger* may be repeated many times before a program even runs, let alone works and gives correct results (Figure 16.1).

Graphics introduces yet another level of debugging. Graphic debugging is assisted by the GKS-C language binding and by the error control built into the GKS standards, which can echo function names and the values passed to them on an ongoing basis. The actual graphics created are also useful debugging tools. Blank screens often mean an error in the transformational geometry involved, while errors with primitives and attributes can generate graphics better suited to the walls of the Museum of Modern Art than to cartography.

Once a program has been written and debugged, it is time to return to the documentation and program purpose to check and verify the program. First, the program should be tested with a small test data set for which the results are known. Few such data sets exist in cartography, so checking with hand calculators is often in order. Attention should be given to special case exceptions to computations, such as taking the square root of negative values or testing for the intersection of vertical, parallel lines.

Once small data sets check out, a real-world cartographic data set should be used. Special cases should be deliberately introduced into the data to check circumstances that may not have been foreseen during design. Real data has a habit of introducing these special cases as well as other limitations, such as size restrictions, into the program and testing the program's design limitations.

Failure of the program at any of these stages may demand going back to a module to make corrections. It may even be necessary to redesign a whole module, which is possible

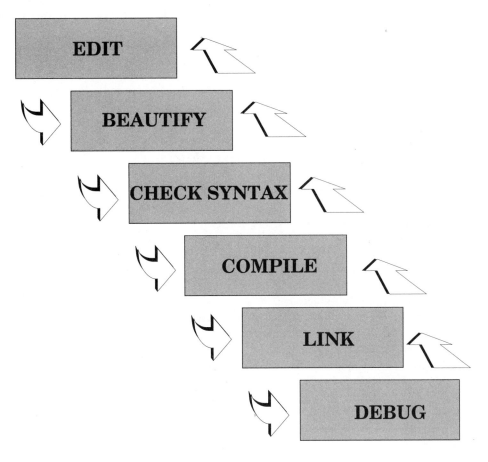

Figure 16.1 The programming sequence.

only with structured programs. If, at this stage, it is found that the entire program structure is poor, this is a good time to give up, as no number of program "patches" can repair a poor overall structure or logic.

As a final stage, many programs are *optimized*. This involves analyzing, often using other computer programs, each module to see how much time is spent in each function and at each stage in each function. The most time-consuming tasks are reworked first, each optimization making the program run faster or use less memory. Some C compilers and operating environments, such as UNIX, have built-in optimizers and program run-time analyzers (the -O compiler option and the profile program). In other cases, optimization is a painstaking process, with each improvement taking seconds off the run times for the program.

Having given a description of the programming environment, it goes without saying that it is critical to cartographic software. Much time will be wasted if the programmer avoids using the tools available, tools that are now parts of the typical C programming environment. A little time invested in learning the sequence used in Figure 16.1 can save hours of wasted time and can result in much more effective and better constructed computer cartographic software.

The programmer should think first and foremost about the map being produced, then about the specifics of the programming language in use. Above all, the design should be organized and planned before a single line of computer code is written. Writing computer programs and designing them are two entirely different tasks, and the worst place to think about design is inside three levels of open control statements.

16.2 THE GKS C LANGUAGE BINDING

In some earlier chapters, particularly in the discussion of the symbolization transformation, the GKS standard was presented as a means by which cartographers can write computer programs that produce maps. We saw that the standard can be implemented at different levels, with different device support and functionality for different levels. We also saw that the link between programming languages and graphics under the GKS standard is the language binding. The ISO specification of GKS lists standard bindings for GKS for FORTRAN, Ada, and Pascal. Because the C programming language has only recently been revised into an ANSI version, there is as yet no standard C language binding to GKS. This does not mean, however, that none exists, merely that there are some differences in bindings between the products of different software vendors.

Fortunately, the GKS standard is quite general when it comes to the major source of differences, that is , the naming of functions. Function names are listed by English language names rather than GKS function names. Figuring out a function name for a different binding, or even transporting a piece of mapping software between computers or programming languages, often consists therefore of translating the names of functions and assuring that arguments in the lists passed to the functions are consistent. A convention used by some vendors for C functions is to use the full rather than abbreviated function names, replacing blanks with underscores (_). On some computers, especially microcomputers, there is a need to abbreviate due to DOS name limitations or limitations on the linking capabilities of C compilers. In this discussion, the simplest form of the name will be given.

A C language binding can be thought of as a set of predefined C libraries that contain all of the functions necessary to use the standard. The functions break down into those that manage primitives and attributes, those that handle interaction with the user, those that control the workstation, those that perform transformations, those that manage segments and the metafile, those that manage errors, and those that allow the programmer to query GKS to establish any of the above.

In all, this makes several hundred functions, because for every primitive, attribute, and interaction function there is a corresponding query function. For example, the function `set_fill_area_color_index()` has a parallel GKS function `inquire_-fill_area_color_index()`. There is a formal structure for the use of GKS functions. The sequence applies for all GKS levels and implementations. The first GKS operation that a program must perform is to open GKS. This performs an initialization of the software on this system for this application. The function used is `open_gks()`. This function sometimes uses as a parameter the name of a file that GKS will use to write GKS error messages. The file should be opened in write mode by the user's program and the value passed should be a pointer to type `FILE`.

The next stage is to open the workstation. This is accomplished using the function `open_workstation()`. Arguments should include an integer identifier for the workstation to be used in successive workstation calls, and a unique number or text string that identifies the type of workstation to GKS. This number is provided with the GKS language binding, and sometimes with the specific device driver installed.

Examples from Sun Soft's SunGKS are `ws_type` values of `"postscript"`, `"hpgl"`, and `"xgl_tool"` for a PostScript, HPGL, and a generic X-windows window respectively. Once opened, the workstation should be activated, or made ready for interaction, using the function `activate_workstation()`. This function simply uses the workstation identifier given to `open_workstation()`.

The "digital canvas" is now ready. The programmer can establish the normalization transformation(s), read any data required, and set any attributes, and is then ready to draw. If no segments need to be defined, the primitives are simply included in the order used. To draw and save primitives in a segment, the function call `open_segment()` is used to create a new segment, which includes all primitives and their attributes until the `close_segment()` function is used. Once defined, segments can be redrawn, renamed, copied between workstations, or written into the metafile. To redraw the entire map, the function call `redraw_all_segments()` could be used.

The way back out of the GKS system is the reverse of the way in. After all segments have been closed, the workstation can be cleared using `clear_workstation()`, deactivated using the GKS function call `deactivate_workstation()`, and closed using `close_workstation()`, and finally GKS can be shut down using `close_gks()`. The GKS system also provides an immediate error exit from GKS so that the programmer can escape from errors elegantly without exiting the graphics levels, using `emergency_close_gks()`. The sequence is illustrated in Function 16.1.

Few differences exist between bindings, with one exception. Many of the C language bindings use single-variable parameters as function arguments. This method follows closely the FORTRAN standard. A line, for example, could be sent to a line drawing (polyline) function using

```
polyline (number_of_points, &x[0], &y[0]);
```

This provides the function with the number of points in the vectors x and y, and the starting addresses of the points. Alternatively, at least one major GKS binding uses a structure of type POINT and instead passes the address of the entire structure to the GKS function.

Each GKS C binding implementation has its own details. Usually, there is a large header file, often called `gks.h`, which contains all the values of the function arguments in a definition header file, so that, for example, color number 6 on a specific device is predefined as

```
#define RED 6
```

Function 16.1

```
/* Program gksbasic : Demonstrate GKS open/close
/* kcc 12-93 Function 16.1
*/
#include <gks/ansicgks.h>
#include "cart_obj.h"
main()
{
        Gchar *conn = NULL;
        Gchar *wstype = "postscript";
        Gws ws = 1;

        /* Open up GKS */
        if (gopengks(stdout, GMEMORY))
              exit();

        /* Open the workstation */
        gopenws(ws, conn, wstype);

        /* Activate the workstation */
        gactivatews(ws);

        /*
         * Generate a map
         */

        /* Deactivate the workstation */
        gdeactivatews(ws);
        /* Close the workstation */
        gclosews(ws);
        /* Close down GKS */
        gclosegks();
}
```

Also, different implementations split the libraries up differently. Sometimes the language binding is separate from the main GKS library. This must be taken into account when compiling and linking C programs.

New GKS implementations, language bindings, and device drivers become available frequently. Trade journals, the so-called glossies, and graphics conventions are good sources of information about these changes. In addition, several GKS implementations are now appearing as shareware. Although these versions are unlikely to support the more exotic computers and graphics devices, the more common ones may be found.

In addition, several companies now make add-on software packages built over the GKS. In a few cases, these systems support advanced mapping functions. Other GKS implementations are very low level (0b, for example), do not support many interactive functions, and allow output only to the metafile. These versions come with metafile translators, which convert the metafiles into the device-specific graphics calls of a great number of different graphics devices. Among the most popular are translations to Post-Script for output on laser-jet printers or to protocols of desktop publishing packages, allowing the output from computer programs to be pasted directly into a paper, article, set of documentation, or even a textbook. Other packages, for example CorelDraw! and IslandDraw, allow metafiles in CGM format to be opened for use and editing.

16.3 WRITING YOUR OWN MAPPING PROGRAM

Assuming that you have read much of this book, have read about the C programming language, understand cartographic data structures, and have gone through the software design process described above, how do you actually go about writing a computer program to generate a map? During the design process, you have asked yourself what kind of map you wish to produce. You have blocked out a set of modules, which can do any necessary input, structure the data, perform any necessary transformations, and then generate the map. The next and most important step is to write the program.

First, you should carefully research the various C compilers available to you. The newsstand microcomputer journals contain large amounts of information to assist in deciding which compiler to buy. If you have access to a larger computer, see if the C language is available and how it is supported. If the computer runs the UNIX operating system, then many of the tools available as part of the programming environment discussed above are available. UNIX can also run on many microcomputers and is the most common workstation operating environment.

Do not start writing programs until you are familiar with the language. When you do, make lots of deliberate mistakes and see how the various parts of the programming environment react to them. If C is your first language, then as mentioned in Chapter 15, you should seek out instruction from a computer scientist. If you are coming to C from Pascal or a similar language, you should have very few problems, but you should find an appropriate C language tutorial and work carefully through it. Pascal programmers should not be tempted to rewrite C to look like Pascal (although it is possible). Although many of the concepts from languages like FORTRAN will help in the conversion, it is better to learn the structured approach from scratch.

A feature that eventually becomes the source of many errors in C programs is the distinction between the types of variables. The C language supports integers (unsigned, short, regular, and long), characters, floats, and doubles, and although variables must be assigned types exclusively, type conversions are not automatic. For example, to make a floating point (real) equivalent of the integer 5, it is not enough simply to assign the value. The type should be explicit in the statement also. This is illustrated in the Function 16.2

Function 16.2

```
/*  ==========================================
/*  Remember to explicitly perform type
/*  conversions in C, otherwise beware!
/*  ==================================== */
#include <stdio.h>
main() {
      int an_integer = 5;
      float a_float;
      printf("Integer value is %d\n",an_integer);
      a_float = an_integer / 2;
      printf("Halved without type conversion the value
            is %f\n",a_float);
      a_float = (float) an_integer / (float) 2;
      printf("Halved with correct type conversion the
            value is %f\n",a_float);
      }
```

One of the best ways to learn about computer cartographic programming is to study the programs written by other cartographers. The list of available sources is short, but it is growing steadily. Cartographers have been slow to publish their computer programs in a form suitable for use by others. As a result, many cartographers have duplicated the work of others without even having the ability to compare the two results.

The programs presented in this book are not complete programs in themselves, but are building blocks around which more useful computer programs may be constructed. In most cases, writing a computer program using these functions will involve reformatting some existing digital cartographic data, entering the data into a programming data structure appropriate to the task, performing any necessary cartographic transformations, and producing a map. The cartographic transformations may involve implementing an algorithm, perhaps one published with a descriptive paper or journal article.

If the programming is done effectively, each of these tasks can be performed by one or more program modules. Remember that these modules should become your own personal library of cartographic functions, and that, for best effect, they should be shared freely with your friends and any other interested persons.

The best way to find weaknesses and errors and to optimize the code is to release programs to other users. Users will expect information such as a manual, and installation details. Also remember, however, that as a user of computer cartographic software you are under an assumed obligation to report problems, errors, and weaknesses to the author of any piece of software, for without this feedback, no improvement will take place. It is also entirely appropriate to report more favorable items of news!

16.4 CARTOGRAPHIC SOFTWARE AND THE USER INTERFACE

The success or failure of a piece of cartographic software is determined not by innovation, sophistication, or accuracy, but instead by usability. Many excellently written and designed computer programs have been passed over entirely because they lack the elements that make the capabilities of the program accessible to the user. The usability of most software, and this includes computer cartographic software, is therefore strongly determined by the user interface of the program. The user interface is the entire set of means by which the user communicates with software to achieve a specific mapping purpose.

In the early days of computer cartography, most software that produced maps ran in batch mode. The means of interaction with the program was by supplying a set of parameters in a file (or even on punched cards) that were read and acted upon by the software. Information could include the names of files, the number of data points in a file, the precise format of an input record, or the numbers of options listed in the documentation. Although batch processing was capable of generating maps, errors were common, since the map output came at the end of a long sequence of operations which happened quite slowly, often overnight.Even generating a map could take several days, but debugging the software for mapping could take weeks or months.

As interactive computer systems became commonplace, many computer mapping systems moved to interactive processing. The earliest user interface in such an environment was the command-driven program. An example is MicroCAM (Figure 16.02).A command-driven program reads one command from the user at a time, and acts upon it either immediately or when a general "action" command is given. Commands usually involve a keyword, such as PLOT or READ, and are followed by a list of values that supply parameters to the program. For example, to read a grid data file, the command

```
READ "Input_file" 200 300 I
```

could be used to read a grid of integers from file `Input_file` with 200 rows and 300 columns. Note that these commands can be placed together into a file and submitted together in batch mode, implying that command mode contains a batch mode.

The next level of sophistication in terms of the user interface is to provide a menu in which commands can be selected by number or letter. Figure 16.3 shows a typical menu for a computer mapping program. Using a menu, selections can be chosen by entering a number instead of the name of the option, in the same way that you can order from a Chinese menu by asking simply for "number 6" instead of La Zi Ji Ding. Many menu systems also allow the name of the command to be typed, thus supporting command lines also.

Increasingly, operating systems are taking advantage of the ability of computer programs to use windows. Windows are areas of the screen where interaction can take place or where graphics can be generated. A windowing system allows the windows to be moved, resized, or selected to be visible, often with the use of a mouse (Figure 16.4).

Selections can be made by moving the mouse to a header and pulling down the menu, or the next menu level, and by holding the mouse button down and moving the mouse. Menus can also appear on the screen at the appropriate time during the execution of the

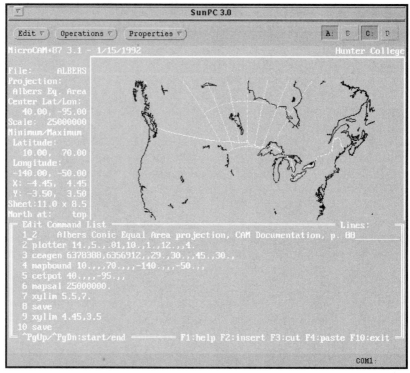

Figure 16.2 A command-line driven software package (MicroCam).

{Terrain_Mapping}

TRANSFORMATION OPTIONS

```
m next menu
w write array to a file
f filter array data
i isopach
n invert array
r resample array
s subsample array
a slope/aspect
h analytical hill shading
t linear trend surface
c Convert feet to meters
u Unravel an array
d Double an array
q Quit


** Select option number :t
```

Figure 16.3 A menu-driven software package (Terrapin: See the companion disk).

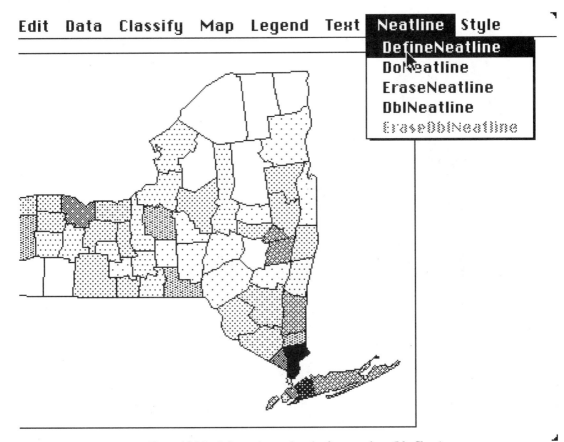

Figure 16.4 A window and menu-based software package (MacChoro).

program, the so-called pop-up menu. The windows environment has been popularized by microcomputers, especially Apple's Macintosh and Microsoft's Windows.

Probably the highest level of sophistication is achieved when mapping software uses all the capabilities of the Graphical User Interface running on top of a computer's operating system. In this case, the program can "inherit" the properties of the GUI, such as window colors, button, and menu formats. This allows all of the capabilities of the GUI to be used, including windows,icons, menus, buttons, dials, and pointers. Examples of this approach are Sun's SUNOS, Solaris, X-windows, Motif, Microsoft's Windows NT, and IBM's OS/2.

Because the window functions are callable from computer programs, the programmer has the ability to incorporate programs that use a consistent interface that is easy for all users, because it embodies concepts which are common between computers. Furthermore, many vendors of these operating system GUIs support GKS, so that GKS calls to locator devices make calls to the mouse, and the "workstation" becomes a window such as an Xterm. An example is xv, an X-windows–based program to display and manipulate images (Figure 16.5). At a higher level of sophistication, several graphics languages and

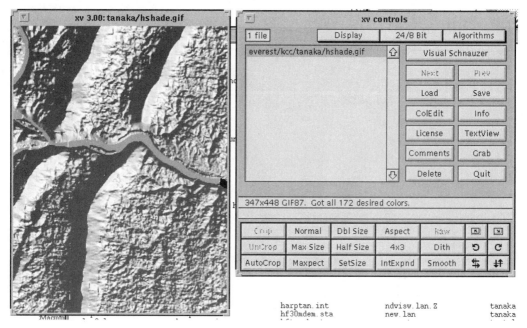

Figure 16.5 A full GUI software package (xv). XV is copyright 1993 by John Bradley.
Used with permission.

toolboxes now support GUI-independent programming, so that a generic program is "compiled" with a "binder" to a particular GUI such as MacIntosh, Windows, or Motif.

For computer cartographic programs, it is important that the means of interaction with the user be kept in mind at all times. Modular and flexible programs make it easy to use one command system during software development and another when the program is finished. A better approach, however, is to write the user interface first. The entire user interaction system can then be debugged, tested, and refined to make the program more "friendly" before the program even works. For this purpose, many large programs are first written with "function stubs," one- or two-line functions that simply report that they were called in the right place and with the correct arguments. Producing a "skeleton" program in this way makes working on a large program much easier and also allows several people to work on a program at the same time.

The user interface is the last challenge facing the analytical and computer cartographer. As we have seen, the cartographer of today must have an understanding of the hardware and software tools of computer cartography. He or she must understand cartographic data, cartographic data structures, and the programming mechanisms by which they can be manipulated. The analytical cartographer must understand cartographic transformations and be able to use the power of computer programming and the programming environment to achieve a higher understanding of maps and mapping.

It is the user interface, however, that in the future will determine how computer mapping systems relate to the vast number of users who make up the rest of the world. If cartography is to move away from being a discipline in which a practitioner disappeared for a time and then delivered a map, if cartography is to become democratic enough to place the production of maps into the hands of the noncartographer, then we must begin by writing mapping software that is usable. This means effective user interfaces and meaningful assistance with decision-making, and it means that much well-written software remains to be produced.

The opportunities presented by the wealth of digital cartographic data now becoming available are extraordinary. How effective this revolution in cartography will be in solving the problems of our world, however, depends on a new generation of analytical and computer cartographers.

16.5 REFERENCES

Enderle, G., K. Kansey and G. Pfaff (1984). *Computer Graphics Programming: GKS--the Graphics Standard.* Symbolic Computation, New York: Springer-Verlag.

Foley, J. D., and A. VanDam (1982). *Fundamentals of Interactive Computer Graphics.* Reading, MA: Addison-Wesley.

Hearn, D., and M. P. Baker (1986). *Computer Graphics.* Englewood Cliffs, NJ: Prentice Hall.

Johnson, N. (1987). *Advanced Graphics in C: Programming and Techniques.* Berkeley, CA: Osborne McGraw-Hill.

Myler, H. R., and A. R. Weeks (1993). *Computer Imaging Recipes in C.* Englewood Cliffs, NJ: Prentice Hall.

Kernighan, B. W., and D. M. Richie (1988). *The C Programming Language.* 2d ed. Prentice-Hall Software Series, Englewood Cliffs, NJ: Prentice-Hall.

Lamb, D. A. (1988). *Software Engineering: Planning for Change.* Englewood Cliffs, NJ: Prentice Hall.

Press, W. H., B. P. Flannery, S. A. Teukolsky, and W. T. Vetterling (1988). *Numerical Recipes in C: The Art of Scientific Computing.* New York: Cambridge University Press.

Appendix

Files on the Companion Disk

The following is a listing of the README file contained on the companion disk. The disk is ASCII format IBM-PC compatible 3.5 inches in size. Files are compressed using PKZ-IP, and a copy of PKUNZIP is contained on the disk. Instructions for decompressing are on the disk. Two compressed files are on the disk. The first contains the C program code as listed in this appendix.This file willl uncompress into the directory structure listed below. The second contains a sample data set in USGS DEM format for the McCall, Idaho, 7.5 minute quadrangle. This data set can be read with read_dem, and analyzed with terrapin.

```
The files on this disk correspond to the example programs in
Clarke, K. C. (1995) "Analytical and Computer Cartography",
Second Edition, Prentice Hall. They are distributed with the
book for the purpose of instruction only. No guarantee is
offered or implied. The disk contents are:

General/ A program to read and generalize strings by
 n-th point retention
 473 Jul 6 11:22 Main.c
 385 Jul 6 10:21 Makefile
2257 Jul 6 11:27 drawstr.c
 33 Jul 6 11:21 extern.h
1045 Jul 6 11:23 general.c
1087 Jul 6 17:27 readstr.c
```

Grid/ Read a grid, convert it to a digital image, and
display using GKS.
```
 419 Jul 14 14:48 Main.c
 388 Jul 14 14:02 Makefile
1300 Jul 14 15:10 display.c
 855 Jul 14 14:59 grdimage.c
 134 Jul 14 14:46 header.h
1972 Jul 14 15:09 init_gks.c
 807 Jul 14 14:59 readgrd.c
```

Invdist/ Read a scatter of (x,y,z) data points and grid
then using inverse distance weighting
```
 239 Jul 8 15:33 Makefile
 552 Jul 12 13:16 confirm.c
 440 Jul 12 13:14 define.h
 164 Jul 8 15:28 edges.c
 475 Jul 8 15:28 extern.h
2029 Jul 8 15:32 grid.c
 255 Jul 8 15:38 header.h
 872 Jul 8 15:30 writedata.c
```

Line/ A Map projection routine
```
 359 Jul 14 11:11 Makefile
2913 Jul 14 10:25 graticul.c
 292 Jul 14 10:21 header.h
1600 Jul 14 10:27 mercator.c
 175 Jul 14 10:05 random.c
1531 Jul 14 10:29 sinusoid.c
6595 Jul 14 11:25 world.c
```

Misc/ Various functions and sample programs from the
book. Some non-programs
```
 621 Jul 14 11:28 example.c
 307 Jul 14 11:30 funct11.03
1340 Jul 14 12:20 funct11.09
 511 Jul 14 12:16 funct16.01
 531 Jul 14 12:20 funct16.02
 293 Jul 14 12:25 funct7.01
 420 Jul 14 12:26 funct7.05
 652 Jul 14 12:24 funct7.07
 470 Jul 14 12:27 funct7.08
 881 Jul 14 12:29 funct7.09
 343 Jul 14 12:22 funct8.01
```

```
 488 Jul  8 14:37 gksbasic.c
1945 Jul  8 15:11 init_gks.c
3006 Jul 14 12:09 intersect.c
1254 Jul 14 12:13 pt_in_p.c
```

```
Near/ Nearest neighbor statistic for points
 450 Jul  8 14:24 Main.c
 231 Jul  8 12:00 Makefile
 369 Jul  8 14:30 distance.c
  21 Jul  8 11:52 extern.h
1367 Jul  8 14:32 nearest.c
 533 Jul  8 14:08 readpts.c
```

```
Polygon/ Various programs for polygons
 896 Jul  8 11:17 Main.c
 237 Jul  8 11:15 Makefile
 520 Jul 14 11:37 area.c
  27 Jul  7 15:00 extern.h
1521 Jul  7 15:12 readpol.c
1090 Jul  7 15:15 readstr.c
```

```
README This file
```

```
Raster/ read a string file and rasterize
 527 Jul  7 15:28 Main.c
 311 Jul  7 16:06 Makefile
1235 Jul  8 10:23 equation.c
 106 Jul  8 10:11 extern.h
 898 Jul  7 15:45 extremes.c
1448 Jul  7 15:48 gridspec.c
1695 Jul  8 10:25 raster.c
1087 Jul  7 15:23 readstr.c
 611 Jul  7 16:06 writegrd.c
```

```
Rubber/ 6 parameter affine rubber sheeting
4387 Jul  8 10:45 rubber.c
```

```
Strings/ Utility programs to read, reformat and display
 string data using GKS
 250 Jul  6 17:20 Main.c
 344 Jul  8 14:46 Make.bin
 353 Jul  8 14:48 Make.draw
 265 Jul  6 17:19 Make.save
2633 Jul  6 17:16 draw_bin.c
2399 Jul  8 14:49 drawstr.c
```

```
  27 Jul  6 17:16 extern.h
1087 Jul  6 17:28 readstr.c
 709 Jul  6 17:25 savestr.c
```

Terrapin/ A set of programs for analyzing terrain
```
 336 Jul 14 15:15 Main.c
 664 Jul 14 15:38 Makefile
 282 Jul  8 15:11 banner.c
2805 Jul  8 15:11 cart_obj.h
 491 Jul 14 15:16 confirm.c
 888 Jul  8 15:11 define.h
2045 Jul  8 15:11 double.c
 407 Jul  8 15:11 exclude.c
 847 Jul  8 15:59 extern.h
1880 Jul  8 15:11 f_to_str.c
3765 Jul  8 15:41 filter.c
 677 Jul 14 15:23 ftmeters.c
 389 Jul  8 15:11 header.h
1367 Jul  8 15:11 i_to_str.c
1098 Jul  8 15:11 invert.c
1160 Jul  8 15:11 isopach.c
 894 Jul  8 15:11 one_i.c
1325 Jul 14 15:20 parameters.c
4000 Jul  8 15:11 pixrect.c
3441 Jul 14 15:41 rddata.c
2019 Jul  8 15:11 rderdas.c
1741 Jul  8 15:11 resample.c
4240 Jul  8 15:11 sa.c
1056 Jul 14 15:20 select.c
3155 Jul 14 15:29 shade.c
1154 Jul  8 15:11 sub.c
 222 Jul  8 15:11 sumsave.c
 724 Jul  8 15:11 swap.c
1683 Jul 12 12:21 transform.c
2190 Jul  8 15:58 trend.c
 952 Jul  8 15:11 unravel.c
5346 Jul  8 15:11 wrbyte.c
1939 Jul 14 15:39 wrdata.c
2464 Jul 11 17:00 wrerdas.c
1227 Jul  8 15:11 wrsurf.c
```

data/ The data sets for all programs
```
 167 Jul 14 09:33 city_points
 117 Jul 14 09:30 city_pops
 450 Jul  8 15:20 everest.doc
```

```
 81173 Jul  8 15:21 everest.pts
   140 Jul  8 10:36 mapimg.pts
426895 Jul  8 10:21 namerica
   246 Jul 14 09:16 point_data
   221 Jul  8 14:34 scatter.pts
  2420 Jul 12 12:12 small.dem
    84 Jul  8 11:18 square.pts
1043754 Jul 14 10:11 world.data
```

headers/ Header file of cartographic data structures
 from the book, see chapter 7.
3285 Jul 14 12:04 cart_obj.h

readdem/ Appendix in the first edition. A program to
 read DEMs from the USGS and write them for
 a terrapin binary read. Second program to
 do the same by Diming Chen of Hunter.

```
  597 Jul  6 17:34 Main.c
  329 Jul 14 15:58 Makefile
  358 Jul  6 17:34 conv_buf.c
  952 Jul  6 17:34 convert.c
  243 Jul  6 17:34 define.h
10535 Jul  6 17:36 dem2lan.c
 4530 Jul  6 17:34 edge.c
  599 Jul  6 17:34 extern.h
  311 Jul  6 17:34 gtval.c
  283 Jul  6 17:34 header.h
 1483 Jul  6 17:34 index.c
  892 Jul  6 17:34 invert.c
  200 Jul  6 17:34 monitor.c
 4311 Jul  6 17:34 rdhead.c
 1384 Jul  6 17:34 read_dem.c
 1381 Jul  6 17:36 stuff.h
  709 Jul  6 17:34 swap.c
  368 Jul  6 17:37 wrhead.c
```

Index

A

Accuracy, 9–10
ACG. See Address coding guide, 76
ADA, 295, 301, 308
Additive color systems, 134
Add-on software packages, 311
Address coding guide, 75, 76
Adjacency, 143
Advanced Very High Resolution Radiometer (NOAA), 95, 97, 99, 109, 113
Affine transformations, 40, 70, 209–11
    depicted, 211. See also Map transformations
Age before present, 174
Aggregate spatial ofjects, 127–28
Air photos: scanned, 99
Album of Map Projections (Snyder and Voxland), 192
ALGOL, 296
Algorithms
    for cartographic transformations, 226
    for distance computation, 194
    for distance minimization, 195
    image-based, 130
    and intersection point of two lines, 198–203
    intervisibility, 269
    and map production, 6
    for point-in-polygon testing, 207, 208
    rasterizing, 238, 239
    thinning, 239
    transformations and, 181
Altek Apache digitizing cursor, 74
American Cartographer, 27. See also *Cartography and Geographic Information Systems*
American Congress on Surveying and Mapping, 43, 112
American Digital Cartography, 40
American National Standards Institute (ANSI), 42, 296, 301
American Research Libraries, 105

American Standard Code for Information Interchange (ASCII), 85, 86, 90, 91
Anaglyphs, 273, 274
Analysis: of map information, 291
Analytical and Computer Cartography, 1st ed. (Clarke), 43
Analytical cartography, 1, 4, 13, 14, 15, 49, 52, 153, 157, 169, 181, 226
    cartographic transformations, 169, 171–73, 223, 228, 233, 274
    and C language structures, 131
    and data structure transformations, 242, 246
    data structure types in, 134–35
    display technologies, 27
    literature of, 11
    and point-to-point transformation, 172, 185, 186, 217
    and polygon overlay problem, 235
    surface models in, 259
    tessellations used in, 144
    and transformation matrix, 184
    and user interface, 316
    and writing software, 304–17
Analytical hill shading, 271–72
Analytical stereoplotter, 249
Anderson, D. P., 268
Animation packages, 38
ANSI. See American National Standards Institute
ANSI-C, 296, 308
ANSI Technical Committee on Computer Graphics, 299
Anstaett, M. R., 140
Antarctica, 10, 64
Anti-aliasing, 237
APL, 296
Apple, 43
    Macintosh, 315, 316
Applications programs, 294
ARC, 91
ArcCAD, 35
Arc files, 164, 165

ARCHIE, 107
ARC/Info (ESRI), 34, 39, 43, 97, 100, 165
Arcs, 137
ARC/View (ESRI), 34, 39, 43
Area: in C language, 118, 121
Area chain, 243
    in C language, 127
Area data, 174
Area features, 137
Area objects, 140–42
Area qualitative map, 285
Area-to-area transformations, 233–35
Area-to-point transformations, 234
Areas, 51, 87, 174
    in TIGER terminology, 79
Aronson, P., 165
Artificial intelligence, 8, 168
ASCII. See American Standard Code for Information Interchange
Assembly language, 295
Association for Computing Machinery–Special Interest Group on Graphics, 299
Association of American Geographers, 40, 43
    Microcomputer Specialty Group, 192
Asynchronous thinning algorithm, 239
ATLAS*GIS (Strategic Mapping Incorporated), 34
ATLAS*PRO (Strategic Mapping Incorporated), 39
Attribute codes: of lines, 93
Attribute data
    sequence of development, 159–60
    structures, 133–35, 153, 159–69
    transformations of, 171
Attribute data models, 160–69
    entity-relationship structure, 166
    hierarchical data model, 161–63
    hybrid structures, 168–69